COVID-19 in Italy

As the COVID-19 crisis began to take shape, all eyes were on Italy, the first Western country to attempt a response to the deadly pandemic. For institutional decision makers and average citizens alike, it was a time of deep uncertainty. As scientists struggled to understand the nature of the virus and how it spread, the gradualness with which information became available caused only deeper uncertainty, as did the inevitable disagreements over which protective actions the government should put in place. Despite some initial delay in its response, the Italian government eventually implemented a nationwide lockdown, which helped to control the spread of the disease but simultaneously created unintended consequences for vulnerable populations, such as small business owners, women, the elderly, and workers living paycheck to paycheck.

Drawing on data surveys conducted during the transition between the first lockdown and staged reopening, this book examines people's risk perception and their willingness to trust the sources and channels of information that were available to them. It also looks at their attitudes toward the protective behaviors they were asked to adopt and the ways in which their own cultural worldviews impacted their support for pandemic response policies. With remarkable depth and candor, respondents reflected on what a post-COVID-19 Italy might look like, filling out the book with the hopes and fears of real people who stared death in the face and lived to tell about it. The book looks ahead to possibilities for future research, policy, and practice. *COVID-19 in Italy* elaborates and tests several aspects of the Protective Action Decision Model (PADM) in the Italian context, introducing the concept of ontological security and insecurity as an explanatory change factor to help interpret the Italian experience of responding to COVID-19.

Lucia Velotti, Assistant Professor, John Jay College of Criminal Justice, Department of Security, Fire, and Emergency Management, City University of New York (CUNY), New York City, NY.

Gabriella Punziano, Assistant Professor, Department of Social Sciences, University of Naples Federico II, Naples, Italy.

Felice Addeo, Associate Professor, Department of Political Science and Communication, University of Salerno, Salerno, Italy.

COVID-19 in Italy

Social Behavior and Governmental Policies

Lucia Velotti, Gabriella Punziano and Felice Addeo

Routledge
Taylor & Francis Group

LONDON AND NEW YORK

First published 2022
by Routledge
4 Park Square, Milton Park, Abingdon, Oxon OX14 4RN

and by Routledge
605 Third Avenue, New York, NY 10158

Routledge is an imprint of the Taylor & Francis Group, an informa business

British Library Cataloguing-in-Publication Data
A catalogue record for this book is available from the British Library

Library of Congress Cataloging-in-Publication Data
A catalog record for this book has been requested

ISBN: 978-1-032-03519-2 (hbk)
ISBN: 978-1-032-03526-0 (pbk)
ISBN: 978-1-003-18775-2 (ebk)

DOI: 10.4324/9781003187752

Typeset in Times New Roman
by Apex CoVantage, LLC

Contents

Figures

Tables

Preface

Social research as an antidote to COVID-19

The current COVID-19 crisis will remain engraved in our collective memory and in that of future generations as a disruptive event, a generational watershed – like the Second World War, the Cold War, or 9/11, to name but a few of the events that shaped a generation and influenced the imagination and conscience of future generations. In fact, the idea that what we are experiencing is not just a period of transient crisis, but a true historical epoch in the making, is constantly evoked in the media and in everyday life with expressions such as "in times of COVID-19" or "in times of pandemic." Thus, it can be said that attempts have already begun to historicize the pandemic, describing and, above all, experiencing it as an event of considerable and profound historical significance. This cultural climate reverberates through the activities of researchers from all disciplines; in fact, scholars have to deal with the study of a phenomenon that is constantly changing, in both its concrete clinical manifestations – think of the spread of variants of the virus – and its possible impacts on the economic, political, and social levels.

Although the first results of the vaccination campaigns are now beginning to show cautious signs of optimism, research on COVID-19 is not standing still, in both terms of public health (new vaccines, more effective treatments, improved health structures devoted to emergency management) and economic and political terms, as many scholars are attempting to create models to predict, with all the epistemological caution, the effects of COVID-19 on national economies in the medium and long term. To this end, the full range of the social sciences – their research objects and research questions, their analytical frameworks, and their research methods – is called into action to understand the pandemic as a social phenomenon and to generate knowledge-based responses that can be useful to institutions and citizens.

Not surprisingly, Sir Anthony Giddens has argued that the COVID-19 pandemic is not just a repetition of previous epidemics; it is something really different, and this difference has its origins in deep structural changes that have helped cause it but also are serving to reshape it. In fact, starting from the consideration that the digital age has radically transformed society, culture, economics, politics, culture, and daily life, he believes we are experiencing the world's first "digidemic." In other words, the current circumstances brought about by the pandemic, although unprecedented, have been profoundly shaped by persistent problems in contemporary societies, such as deep-rooted racial and economic inequality, industrial and agricultural production systems that have generated a development that is still far from sustainable, national health systems continually harassed by cutbacks to favor private healthcare, and a public opinion that, because of pervasive social media, has become too humoral and present to the point of conditioning with its anxieties the ability of the world's democracies to deal with major crises.

Thus, COVID-19 undoubtedly disrupted the organization of societies and upset the daily lives of people all over the world. Virtually no dimension of social life has been left untouched. From difficult individual choices about what protection and safety measures to take, to government decisions on public health and economic policy; from managing couple life and family relationships, to the drastic changes that have occurred for many in their workplaces; from the need to rethink, due to social distancing, almost all human interactions on the basis of the socio-technical mediation of techno-logical devices, to the proliferation of cases of misinformation and disinfor-mation, whereby false news or exaggerations, even from official sources, have had a detrimental impact on the management of the pandemic crisis by governmental and health institutions.

From a strictly sociological point of view, the pandemic represents, first and foremost, a great epistemological challenge for the social sciences, because we are talking about something that is not yet a well-defined research object, i.e., ready for a complete posteriori analysis. It is a research object that is difficult to grasp, because it is constantly changing before our eyes. Moreover, the climate of uncertainty caused by the COVID-19 emer-gency undoubtedly had the impact of generating so-called pandemic social practices (Werron & Ringel, 2020), i.e., social practices that emerged and continued during the pandemic and were somehow linked to the discovery and spread of the virus. In general, this conceptualization refers to all kinds of practices that we might call generative, as they not only deal with the virus but also literally give birth to it, i.e., they turn it into a concrete social phenomenon to be reckoned with. For example, many analysts believe that the shock to national economic systems will be greater than that caused by the 2008 economic crisis and, among the likely consequences of the pandemic, there will be inevitable cuts in public spending on education and research that will have the unfortunate consequence of exacerbating existing inequalities. Catastrophic events, such as a pandemic, might increase social and health inequalities, but they can also be an opportunity, as Amartya Sen argues, to bring about progress in today's societies in terms of greater social cohesion and improved overall population health.

Social research has the duty, along with other scientific activities, of exploring and possibly creating the potential conditions to proceed toward the positive scenario envisaged by Amartya Sen, i.e., to try to ensure that opportunities may emerge from the crisis, through rigorous and in-depth sociological analyses of what is happening and through empirically grounded proposals aimed at avoiding, or at least mitigating, the negative effects of the pandemic crisis. Unfortunately, social research is currently suffering the consequences of the restrictions and limitations induced by the pandemic's containment measures. First of all, despite the undoubted practical advantages of being able to organize meetings remotely, the persistence of these forms of academic and scientific debate, carried out exclusively online, is exhausting and alienating researchers, and not a few believe that the absence of physical forms of cultural and scientific exchange (conventions, seminars, conferences) could seriously jeopardize the future of social research.

Yet, the most serious problem relates to concrete empirical research activities: Traditional social science research methods are either impractical or require adaptation to protect the health and safety of both researchers and research subjects. COVID-19 has made all forms of data collection based on personal contact unfeasible, and this has been particularly detrimental to qualitative approaches, as most applied research methods are based on interactions between subjects within a social and cultural context. Research carried out using quantitative approaches also had to adapt to the new conditions created by the COVID-19 emergency, leading to a mass exodus of quantitative research in favor of online surveys. Over the past few months, our email inboxes and social profiles have been literally flooded with requests to participate in surveys carried out by universities, research institutes, market research companies, and ordinary citizens – undergraduates, for example. The scientific validity of these surveys is obviously fluctuating, and it is likely that the current modus operandi, being acquired and internalized by younger researchers in particular, could significantly lower the average quality of future research work.

However, this emergency situation paradoxically also had positive side effects, especially in accelerating the adoption and diffusion of digital research practices, i.e., so-called digital methods. In other words, as with many other aspects of social life, the pandemic has acted as an accelerator of practices gradually but also very slowly, establishing that COVID-19 has overturned the inertia of digital innovation in social research.

The possibility of being always connected in a situation of confinement or mobility restriction not only caused an exponential increase in the production of big data but also facilitated and intensified the production of small data, through the use of traditional research methods declining in their online version.

Social research has thus begun to carve out a crucial role in understanding the social and cultural effects of COVID-19, using digital methods to explore and interpret pandemic practices, understood as both concrete actions to avoid contracting and spreading the infection (social distancing, the adoption of individual protection tools such as masks, respect for lockdowns) and cultural and communicative practices to adapt to the restriction of the possibility of moving and having social relations (smart working, distance learning, online goods, and purchase of services).

It is, therefore, clear that the future of social research lies in its ability to hybridize with digital technology without losing its nature and distinctive features. Paradoxically, COVID-19 has proven to be a great opportunity for social research to embark on a path of theoretical reflection and methodological innovation.

Social research could, together with other scientific activities, be an essential tool to be used in crisis situations, such as the one created by COIVD-19. Social researchers can help policy-makers develop and implement shared actions and strategies that can lead to inclusive and sustainable recovery after times of crisis. Moreover, social research can ensure that the voices of those communities in distress are heard and discussed, and that those affected can be involved in decisions that affect them. Thus, the social sciences are regaining a central role within the public debate on society that seems to have taken place in recent decades.

This book offers an excellent example of how social research can address a complex and evolving phenomenon such as the current pandemic, providing a theoretically and empirically grounded analysis.

The theoretical framework used by the authors, based on Lindell and Perry's Protective Action Decision Making (PADM) model and on Giddens' concept of ontological (in)security, is very effective in analyzing a complex phenomenon such as the current one and providing a broad interpretative framework to understand the individual and collective actions through which the problems caused by the pandemic are addressed.

From a methodological point of view, the research design presented in Chapter 3 is clear and straightforward and a good testimony to not only how COVID-19 has impacted social research but also the authors' ability to adapt to the situation. For example, the two-stage sampling procedure, using social media as collector points, proved to be very effective, leading to the achievement of a good sample size in a short time and with adequate heterogeneity of respondents. Moreover, the authors' analytical work is refined but is also presented in a way that is understandable even for the reader not familiar with statistical analysis. Indeed, the combination of multivariate analysis techniques adopted by the authors has as its main result the creation of effective typologies with which to interpret the behavior of the individuals interviewed.

In conclusion, it is to be hoped that within the scientific production of the social sciences, more and more space will be found for works of this kind, which bear concrete witness to the fact that social research is an antidote capable of providing the necessary antibodies to deal with critical situations such as the one we are currently experiencing.

Enrica Amaturo

Reference

Werron, T., & Ringel, L. (2020). Pandemic practices, part one. how to turn "Living Through the COVID-19 Pandemic" into a heuristic tool for sociological theorizing. *Sociologica, 14*(2), 55–72.

Acknowledgments

This book could not have been possible without the excitement, vision, and commitment of the many people that have contributed to it, each in his or her own important way. We would like to thank these people in order of when they contributed to this project.

First of all, a special thanks goes to our survey respondents, who took the time to participate in our study. We were surprised by the outpouring of interest in this survey, and by the depth of the comments left in the open-ended section of our survey instrument. This study would not have been possible without them. We hope that this book, with its suggestions, will contribute to the resilience of our communities.

Second, our thanks go out to our daughters, Dhana Maria Rita, Elena Francesca, and Alice, who also contributed to the realization of this book through their patience while we carried out our research and writing. Many times you were asleep in our arms during long meetings, and many other times you cheered us up with your smiles and your inspiring way of looking at the world. Thank you, babies; we love you immensely.

Third, we would like to thank Ms. Faye from Taylor and Francis, who immediately recognized the importance of this book project, as she was living in Italy and experienced firsthand the devastating effects of the pandemic on the Italian population. Thank you for your support. Your excitement was one of our greatest motivators. We are honored to share our research findings within the FOCUS edition of Taylor and Francis.

Fourth, we would like to acknowledge the work of Dr. Albert Liberatore, founder of Edits Made Easy LLC, for his attention to detail in copyediting the manuscript. Dr. Liberatore worked with us in trying to meet our deadlines and providing precious and timely feedback. We also appreciate the human side of Dr. Liberatore and his humor.

Last, but not least, we would like to acknowledge and thank the entire staff of the Office for the Advancement of Research at John Jay College, which provided a grant to fund this work.

Thank you all. We are honored to have worked with you.

Lucia, Gabriella, and Felice
New York City, Napoli, and Salerno
July 30, 2021

Author biographies

Lucia Velotti, PhD, is an assistant professor in Emergency Management and Disaster Science at the John Jay College of Criminal Justice in the Department of Security, Fire, and Emergency Management, City University of New York (CUNY) in New York City. Dr. Velotti's experience in the field of disaster and emergency management is both national and international. She has carried out several field studies in Haiti, Japan, the Netherlands, and Italy, and also in Oklahoma, Alabama, and North Carolina in the U.S., following disasters like earthquakes, tsunamis, tornadoes, and floods. Dr. Velotti has been awarded several grants from the National Science Foundation (NSF); the Disaster Prevention Research Institute (DPRI) of Kyoto, Japan; and City University of New York. Dr. Velotti graduated from the University of Delaware, where she was a research assistant at the world-renowned Disaster Research Center (DRC). She is the co-lead of the Federal Emergency Management (FEMA) Special Interest Group (SIG) on Leadership and Service Learning, and co-chair of the International Research Society Public Management (IRSPM) panel on Emergency Services.

Dr. Velotti is also a member of the following interdisciplinary research groups within Converge COVID-19: Risk Communication in Concurrent Disasters, and Bridging Needs with Research through Action-Oriented Community Design, funded by the National Science Foundation (NSF).

Felice Addeo, PhD, is an associate professor in Social Research Methods at the University of Salerno, Department of Political and Communication Science, in Italy. He is a methodologist with vast experience in social science research. Since 2017, he has served as the director of the Summer School in Social Research Methods (www.paideiascuoleestive.it/) held in Salerno, Italy.

Dr. Addeo is also an independent expert reviewer for the European Commission General Directorate for Research and Innovation, as well as an independent evaluator for the European Commission Research Executive Agency. His current research interests are social research methods, epistemology, quantitative research, qualitative research and mixed methods, online research methods, netnography, sustainable development, social cohesion, and migration studies.

He has published three books, several book chapters, and many papers in a wide range of national and international academic journals, such as *New Media & Society*, the *Italian Journal of Sociology of Education*, *Telecommunications Policy*, and *Sinergie, Mercati e Competitività*.

Gabriella Punziano, PhD, is an assistant professor in Sociology and Methodology in the Department of Social Sciences of the University of Naples Federico II, in Italy. Dr. Punziano is a scientific member of the Converge COVID-19 Working Groups for Public Health and Social Sciences Research, funded by the National Science Foundation (NSF), where she focuses on risk communication in concurrent disasters. She is also a member of the following working groups: the University Tourism Observatory; the Youth Observatory; the Digital Methods and Social Research group; and other national and international research networks mainly focused on methodological developments. Dr. Punziano, during her post-doctoral work at the Gran Sasso Science Institute (GSSI) in L'Aquila, investigated how emergent groups contributed to recovery efforts in the rebuilding of L'Aquila in the aftermath of the 2009 earthquake. Since 2018, she has served as the director of the Paideia Association for Culture and Advanced Training in Human and Social Sciences and as a teacher at the Summer School on Method and Social Research. Her research interests include the methodology of social research; the new analytical frontiers and the challenges introduced by big data, mixed and integrated, and digital perspectives; social policies and welfare regimes concerning social inclusion, territorial cohesion, and community integration; the analysis of public, institutional, and political communication phenomena through innovative content analysis techniques; and risk communication analysis on social and digital platforms.

1 COVID-19 pandemic in Italy

An overview

Introduction

As of January 8, 2021, the number of deaths caused by the COVID-19 pandemic has reached 1.9 million worldwide (Johns Hopkins University, January 8, 2021), and it is unclear when people will gain herd immunity. This terminal stage of a pandemic is achieved when a large enough segment of the population is immune to the disease, making it unlikely that it will continue to spread from one person to another.

In order to discuss the unfolding of the events related to the COVID-19 crisis in Italy and its management, it is necessary to provide a definition of pandemic. The word "pandemic" comes to us "from Greek *pandemos* 'pertaining to all people' . . . from *pan-* 'all' . . . + *dēmos* 'people'" (Online Etymology Dictionary, n.d.). Thus, a pandemic involves people worldwide; it differs from an epidemic, which affects a more restricted or contained area, such as a community. In essence, a pandemic is a global epidemic. The term "pandemic" does not tell us anything about the severity of the disease, rather indicating the extent of its diffusion; the word "pandemic," therefore, refers to a disease's geography.

In addition to this contagion geography, another element that needs to be considered when defining a pandemic is the disease's novelty. The World Health Organization (WHO) defines a pandemic as "the worldwide spread of a new disease" (WHO, 2010). The novelty of the disease is crucial because it deepens the uncertainty faced by the affected communities.

A pandemic in disaster science and crisis management can be described as a *transboundary crisis* (Ansell et al., 2010; Boin, 2019; Goldin & Mariathasan, 2014; Quarantelli et al., 2006) or a *catastrophe* (Quarantelli, 2000, 2006). In a catastrophe, it is most likely that neighboring communities cannot help one another and, instead, compete for scarce resources (Quarantelli, 2000, 2006). Unlike emergencies and disasters, catastrophes make the impacted communities more vulnerable, since they cannot seek help from neighboring

DOI: 10.4324/9781003187752-1

communities; there is a loss or a lack of both facilities and response personnel. Moreover, while all disasters are local, insofar as they always impact local communities, catastrophes are distinguished by the fact that they demand attention from the national government (Quarantelli, 2000; Quarantelli et al., 2006). The definition of COVID-19 as a transboundary crisis provides a better idea of the event's level of management and governance. Transboundary crises, as the words indicate, do not have boundaries, whether legal, political, or geographical (Boin, 2019). Pandemics are well-established examples of transboundary crises (Baekkeskov, 2015).

This first chapter will provide an overview of crisis management, with a focus on the management of COVID-19 in Italy. The chapter discusses some of the elements related to the Italian case that make it difficult to manage transboundary crises (Boin, 2019), such as:

1. The existence of multiple domains and multiple manifestations: Transboundary crises may involve several countries;
2. The incubation and rapid escalation: The level of development of transboundary crises varies – they can rapidly escalate or be reabsorbed, and then can explode again;
3. The difficulty of pinpointing where a crisis started and how, exactly, it evolved;
4. The involvement of multiple actors with conflicting responsibilities: Transboundary crises can generate crises of governance and leadership; and
5. The lack of ready-made solutions: Transboundary crises require non-routine responses.

In this chapter, we will see the characteristics of the novel coronavirus, COVID-19, and how it compares to other pandemics, followed by a discussion of the elements of crisis management and governance. Next, we will move on to a timeline of the Italian management of COVID-19, describing how events unfolded and highlighting the deep uncertainty that decision makers and laypeople alike encountered in handling the situation. As one of the first Western countries to respond to COVID-19, the Italian government faced uncertainties about which protective actions it ought to recommend to contain and mitigate the effects of the virus. In the initial stage of the crisis, there was deep uncertainty regarding the modalities through which the virus spreads, how long it stays on surfaces, and how people get infected. Exacerbating the problem was a crisis of governance and leadership that was generated by the conflicting involvement of the central and regional levels of the Italian government in organizing the response and recovery effort, and we will examine this in some detail. The lack of ready-made solutions will be clearly illustrated by a brief excursus on the policies that were enacted by the government, coupled with people's behaviors as they sought to adapt to a new way of living. The chapter will conclude with an overview of the book's organization and structure.

COVID-19 and previous pandemics

Pandemics are not a new threat to humanity. The earliest report of a pandemic is Athens's plague (430 BC), which killed approximately 25% of Athens's population and whose origins are still unclear (Leonard, 2020). Understanding and comparing previous pandemics is helpful because it provides an opportunity for emergency managers and institutions seeking containment to reduce the uncertainty over how to respond, mitigate, and recover from a pandemic event, provided that similarities can be found. Pandemics can be compared to each other based on their origin, duration, disease diffusion, mortality rate, vulnerable populations, and historical context. However, comparison is not always easy since pandemic events are loosely defined, and therefore, there is not always agreement on which events to consider as pandemics. For instance, the WHO, the main organization in charge of declaring an event's pandemic status, declared HIV an epidemic of global proportion, but not a pandemic.

The WHO loosely defines pandemics as "the worldwide spread of a new disease" (WHO, 2010). Elevating an infectious disease to pandemic status is not something to be taken lightly, as it might have global financial and social repercussions. The WHO system declares that a virus has reached pandemic status when there is a sustained and widespread diffusion of the virus to humans, based on six pandemic alert phases (WHO, 2009). Phases 1 through 3 signal the virus's diffusion primarily to animals and to some human beings. Phase 4 concerns sustained human transmission. And Phases 5 and 6 occur when there is widespread human transmission. These phases of alert should guide governments internationally to plan and act accordingly. However, the Council of Europe memorandum, issued in June of 2010, reported criticism toward the WHO for declaring H1N1 (swine flu) a pandemic without disclosing potential conflicts of interest (O'Dowd, 2010). Critics objected that the status of the 2009 H1N1 influenza did not justify the pandemic declaration and only created panic among governments and a waste of public resources. In 2010, Wolfgang Wodarg, the head of health at the Council of Europe, stated that raising H1N1 influenza to the status of a pandemic was

> one of the greatest medical scandals of the century. . . . We have had the mild flu – and a false pandemic. . . . In order to promote their patented drugs and vaccines against the flu, pharmaceutical companies have influenced scientists and official agencies, responsible for public health standards, to alarm governments worldwide.
>
> (Macrae, 2010)

A similar situation also occurred for the Western African Ebola epidemic in 2014, when the agency was accused of downplaying the event.

Officially, after the WHO elevated H1N1 influenza (swine flu) to the status of a pandemic, COVID-19 was the second declared pandemic of the twenty-first century. On March 11, 2020, the WHO declared the novel coronavirus (COVID-19) outbreak to be a pandemic, with diffusion to 114 countries, 118,000 cases, and deaths (WHO, 2020). The WHO delayed declaring COVID-19 as a pandemic, even though it had warned the international community about it on more than one occasion. The declaration of COVID-19 as a pandemic was not a decision taken lightly by the WHO, as it was fully aware of the consequences of such a statement on social lives and worldwide economies. The behavior of the WHO surrounding its definition of COVID-19 as a pandemic shows how a pandemic is not an easily identifiable event. In this, pandemics are notably different from other hazardous events, such as earthquakes and hurricanes, which have much clearer identifying characteristics.

From an epidemiological point of view, COVID-19 is a severe acute respiratory syndrome coronavirus (SARS-COV 2) that was first identified in December of 2019 in Wuhan, China. The disease is airborne, meaning that it can be transmitted from an infected person through sneezing, coughing, laughing, and contact with infected surfaces. The 2009 H1N1 (swine flu) was also transmitted through respiratory droplets and infected surfaces. However, H1N1 was not as severe as COVID-19, and it was soon redefined as a mild flu. Symptoms of COVID-19 range from sore throat and fatigue to nausea, vomiting, and diarrhea. However, some people can be asymptomatic. The existence of asymptomatic infection significantly complicates our ability to determine the number of infected people. And this, in turn, can make infection containment almost impossible if it does not involve aggressive testing to find and isolate those who are infected.

The containment of COVID-19, compared to that of the 2013–2014 Ebola disease, is much more difficult. Ebola was more preventable, since it was not airborne; it was transmitted through contact with contaminated blood or bodily fluids. Thus, in the case of Ebola, it was easier for people to identify the threat and avoid contact with it. With COVID-19, though, avoiding infection involves self-distancing and quarantining – much more complicated processes. Beyond this, in comparison with other pandemics, COVID-19 has the longest incubation period: from 2 to 14 days and, in some cases, up to 27 days (Yuko, 2020).

Pandemics should also be compared in terms of the exposed population, the global population, and the moment they occur in history. The 1918 Spanish flu, like COVID-19, was a respiratory virus, and it spread all over the world through the soldiers fighting World War One. The disease probably originated in the USA and was transmitted to humans from pigs. The Spanish flu killed more people than the war did, with an estimated 50 million deaths. The vulnerable population was aged 65 and younger, with a concentration in the age group 20–40 years old (Marantette, 2021). Compared to the Spanish flu outbreak, the world is now more populated; "there were fewer than 2 billion people in 1918, and now there are 7.5 billion" (Mineo, 2020). Another difference is that the population is now more mobile; people did not travel by air in 1918. Today's society is also more complex – due, among other things, to the increased interconnectedness of the supply chain.

If all these factors are not enough, the sustained spreading of COVID-19 is also exacerbated by the variants of the virus. There are already three variants that have been identified worldwide: the English, the Brazilian, and the South African. In particular, the English variant seems to spread more quickly and to be more deadly than other forms of the virus. Other COVID-19 variants recently identified take their names from the Greek alphabet, such as the alpha, beta, gamma, delta, and lambda variants. The delta variant seems to be of greatest concern as of August 2021. In terms of ending the pandemic, there are several vaccines now available, but their effectiveness in general, and their effectiveness against the variants specifically, are not yet fully known. The management of the COVID-19 crisis, and in particular the role of government in its containment, is punctuated by deep uncertainty about the new social and physical environments in which both people and institutions must learn to live and to which they must adapt.

Crisis management and governance

The term "crisis" is loosely used in formal and informal conversation to indicate a disruption of routine for people and organizations. Crises are situations or events "perceived by a community as capable of disrupting its core values or generating a threat to essential facilities" (Boin & 't Hart, 2007) under conditions of deep uncertainty. Based on this definition, it is possible to distinguish three characteristics of crisis management: (1) the presence of a threat, (2) the uncertainty, and (3) the urgency (Boin & 't Hart, 2007; Boin et al., 2005).

The presence of a threat is sometimes difficult to establish since the threat is not always measurable in terms of physical damage or diffusion. For instance, a threat can refer to a society or community value system. COVID-19 threatened several values, such as societal safety, citizens' freedom of movement, solidarity (because of a fear of the other), and people's understanding of the physical world that surrounds them. The second element that characterizes a crisis is the urgency of the need to do something. In essence, urgency is "the perception that time is at a premium: the threat is here, it's real, and it must be dealt with as soon as possible" (Boin et al., 2017, p. 6). The third element of a crisis is uncertainty. Uncertainty refers to the nature and consequences of a threat. For instance, there can be uncertainty about what is happening and how the situation will evolve. In a pandemic, it is fundamental to understand adequately both the threat and how the crisis will unfold. What will the consequences of this crisis be for society? How severe will the crisis be? How can this severity be measured (e.g., number of deaths, number of people who lost their jobs, etc.)? How much is known about the threat? Uncertainty can be mitigated by looking at similar situations that have occurred previously, if there are any.

The management of a crisis is not an easy task. From its initial acknowledgment to its resolution, the cycle of a crisis is comprised of four main stages: sense making, decision making, meaning making, and terminating the crisis – followed, of course, by learning from it (Boin &'t Hart, 2007; Boin et al., 2005). The stage of sense making of the crisis is essential because a crisis cannot be managed if it is not first recognized. At times, signals of an impending crisis are disregarded by public managers. Thus, it is not uncommon to have a situation of crisis denial (Jervis, 2017; Lebow & Stein, 1994; Parker & Stern, 2002; Perrow, 2011). Of course, to recognize and make sense of a crisis, accurate information is needed. The problem is that gaining a complete and timely understanding of the situation is complicated by the fact that information is scattered among many organizations, making it difficult to know what kind of information exists and is available. At this stage, it is helpful to have a way to centralize all the information. The challenge, then, is how to make sense of information, especially when it comes from different sources. In essence, the stage of sense making involves public managers' cognitive abilities, that is, "the capacity to recognize the degree of emerging risk to which a community is exposed and to act on that information" (Comfort, 2007, p. 189).

Once a crisis is acknowledged, it is time to decide how to manage it. Thus, the decision-making stage is about the governance of the crisis. One of the most important decisions to be made is about who has the power, legitimacy, and responsibility to manage the event. The two main options for managing the governance of a crisis are: (1) coordinating among organizations or agencies or (2) centralizing the power within one organization (Boin, 2019). The centralization of powers carries the problem of legitimacy and, therefore, of maintaining political and public support. In the context of transboundary crises, power centralization becomes even more challenging to implement. Power coordination, on the other hand, is troublesome because it requires that all the parties work together with trust in one another. Trust can be difficult to achieve among organizations that have never worked together before.

After deciding which organizations have the legitimate authority to respond to the crisis, it is time to communicate to the public what it will take to solve the crisis. In the meaning-making stage (Ansell et al., 2010), it is imperative to share a common definition of the situation with national and international communities. Creating a shared and legitimate narrative is vital because this narrative will compete for legitimacy with others that can be more or less official. It is not uncommon to encounter groups of so-called negationists, people who deny the trustworthiness of the governmental perspective, or people who claim that the government is plotting against them. Another challenge of communicating with the public is the enhanced likelihood of giving them conflicting messages, largely due to the number of organizations responding to the crisis. Conflicting messages, in turn, undermine the credibility of the message being given.

The last stage of the crisis is that of terminating the crisis and learning from it. Terminating the crisis is about the decision makers' ability to understand when a crisis can be safely ended for all the affected people, with no unintended consequences (Stern, 2013). Ending the crisis may seem like a straightforward task, but it is not. Crises can be ended too soon or too late ('t Hart & Boin, 2001).

Once the crisis has finally ended, learning from it is of the utmost importance. To learn from a crisis, it is necessary to assess what worked well and what needs improvement if future crises are to be managed more effectively. It is critical to remember that the "lesson learned" document produced in the aftermath of a hazardous event is only the starting point, not the endpoint (Stern, 2013).

The Italian response and recovery timeline

The Italian experience with managing the COVID-19 crisis followed a circular pattern of sense making: deciding what to do, trying to resolve the crisis, and starting the cycle again with each new pandemic wave. Crises are events that are not "neatly delineated in time and space" (Rosenthal, 1998). "A crisis may smoulder, flare-up, wind down, flare up again, depending as much on the pattern of physical events" (Boin & 't Hart, 2007, p. 3).

In this section of the chapter, we will consider a timeline of the events that marked the Italian response and recovery effort. This timeline groups events based on the stages of crisis management, as follows:

1. Sense making and decision making of the COVID-19 crisis;
2. Decision making and meaning making: Phase 1 – From regional to national lockdown;
3. Terminating the crisis and learning from it: Phase 2 – Staged reopening;

4. Sense making and decision making: The second pandemic wave and staged lockdowns;
5. Terminating the crisis and learning from it: Phase 3 – The vaccine campaign.

Sense making and decision making of the COVID 19 crisis (January 2020 to February 2020)

The sense-making process of the Italian crisis ran from January to February of 2020. However, sense making is an ongoing aspect of every crisis because of the state of deep uncertainty typical of crises. Thus, "sense making" here primarily means the acknowledgment that a crisis exists. In the sense-making stage of the COVID-19 crisis, Italy went from crisis denial (Jervis, 2017; Lebow & Stein, 1994; Parker & Stern, 2002; Perrow, 2011) to the normalization of risk and initial underreaction (Capano, 2020). "Crisis denial" implies that decision makers felt the situation was under control. Note that the understanding of COVID-19 in terms of symptoms, health consequences, and diffusion was updated several times.

The end of January 2020 represents the starting point for an escalation of events leading to the Italian COVID-19 crisis. On January 30, 2020, two Chinese tourists traveling from northern Italy were hospitalized for COVID-19 in Rome. The following day, the prime minister declared a national state of emergency. Once the national state of emergency was in place, the National Department of Civil Protection took charge of the response under decree 1/2018 (Comfort et al., 2020). The first protective measure issued was the cancellation of all flights to and from China. However, Italian flight cancellations did not impede Chinese people from entering Italy by way of other European countries.

At the inception of the outbreak, wealthy northern Italian regions, such as Lombardy and Veneto, oversaw the response. It is notable that Lombardy and Veneto each took a different approach. The Veneto region isolated all people who could potentially be a threat and invested in swabs to identify and isolate people affected by the virus. Meanwhile, the Lombardy region and the city of Milan, with the campaign #Milanonsiferma (Milan will not close), decided to ignore the presence of COVID-19 cases to keep investors and industries appeased. During this stage, the pandemic was still managed at the regional level. However, it was not long before the national government stepped up and decided to put several northern Italian regions under lockdown. This decision was then extended to the rest of the country. The two different ways of managing the situation by these two separate regions may have been an attempt on the part of regional governors and other leaders to keep regional identities intact. The city of Milan has historically been recognized as a city that never stops working, and as the capital of the industrial and wealthy North.

During February of 2020, intolerance toward Chinese communities fomented. This intolerance was largely instigated by Matteo Salvini, the leader of the Italian political separatist movement Lega Nord (North League). Salvini instrumentalized the pandemic for what appeared to be political reasons, blaming Chinese communities for the spread of the disease.

Decision making and meaning making (March–April 2020): Phase 1 – From regional to national lockdown

The decision-making and meaning-making stage focuses on the ways in which the crisis was managed at the governmental level, together with the ways in which a shared narrative was presented to the public regarding how the pandemic happened, how it spread, and how to contain it.

On March 1, 2020, the Council of Ministries divided the country into three zones (La Repubblica, 2020):

1. *The Red zone* was made up of the municipalities of Bertonico, Casalpusterlengo, Castelgerundo, Castiglione D'Adda, Codogno, Fombio, Maleo, San Fiorano, Somaglia, and Terranova dei Passerini in Lombardy, and the municipality of Vo' in Veneto. The municipalities in the red zone had to quarantine.
2. *The Yellow zone* was comprised of the regions of Veneto, Emilia-Romagna, and Lombardy. In all these regions, gatherings for the purpose of social and sporting events were forbidden. Movie theaters and museums were also closed.
3. The rest of the country followed only a set of safety and prevention measures.

On March 7, 2020, a leak from the Council of Ministers to the press caused some people to panic about the Lombardy region coming under lockdown. News of the imminent lockdown caused some people to rush to the train stations in an attempt to reach their families in southern Italy (Il Messagero, 2020). Just days later, on March 10, the lockdown was extended to the entire nation.

And the next day, March 11, the WHO declared the COVID-19 outbreak a pandemic (WHO, 2020). This same day, Prime Minister Giuseppe Conte decided to tighten the lockdown measures by allowing only pharmacies and food stores to stay open. Exactly one week later, on March 18, 2020, the WHO reported that there were 74,346 laboratory-confirmed cases of COVID-19, and more than half of these were in Italy (Nacoti et al., 2020).

Far from being extreme, the Italian government's institution of a national lockdown resulted from the government's awareness that COVID-19 was developing from an initial regional crisis into a transboundary catastrophe. Thus, the unfolding of the events that led from an initial focus on a few regions to a national lockdown might have shaped Italians' risk perception and protective actions. The decision to put an entire country under lockdown created unintended consequences for vulnerable populations and other entities, such as small businesses, women, the elderly, and vulnerable workers. The government agreed with health professionals and scientists and developed a response plan made up of three phases: Phase 1 – Response; Phase 2 – Short-term recovery; Phase 3 – Long-term recovery (vaccine).

During Phase 1, despite some initial delay in responding to the emergency, the Italian national government's decision to put the entire country under lockdown was a desperate attempt to slow the virus's spread ("flatten the curve"). One of the driving factors behind the lockdown was the fear that if the virus spread too quickly, it would overwhelm Italy's public healthcare system, which was already burdened by a lack of capacity because of previous socio-structural vulnerabilities. In this stage of the response, the message conveyed to the population was clear: Everybody had to stay at home (Decreto del Presidente del Consiglio dei Ministri, March 9, 2020). On April 10, the president of the Council of Ministers, Prime Minister Giuseppe Conte, signed a decree extending the lockdown to May 3. The decree also stated that, beginning on April 14, bookstores, stationery stores, and stores for children's clothes would be reopened. At the end of April, the central government enacted a Phase-2 decree (DPCM, April 26, 2020), detailing the rules for a staged reopening that would begin on May 4.

Terminating the crisis and learning from it (May–September 2020): Phase 2 – Staged reopening

The second Decreto del Presidente del Consiglio dei Ministri (DPCM, April 26, 2020) announced the decision to terminate the crisis and loosen restrictions. This decision was based on the rate of contagion. The decree of May 4, 2020, marked the beginning of Phase 2, which lasted until June 3, 2020. This second phase consisted of an initial reopening of manufacturing industries and construction sites. The Phase-2 reopening was determined on the basis of activities, not based on an index of infection or contagion. Schools, bars, restaurants, and hair salons would stay closed until May 18. Then, on June 3, international travel and inter-regional mobility within Italy were restored.

The month of June was significant since the government sent signs of recovery and the entire country seemed to be optimistic about returning to normal. On June 9, the committee of experts appointed by the central government released the "Iniziative per il rilancio 'Italia: 2020–2022,'" (Comitato di Esperti in Materia Economica e Sociale, 2020), a strategic document laying out Italy's directions to be followed in an effort to become more resilient in the face of future crises. At this time, the hospitality industry was discussing strategies for reopening in a way that would ensure customers' safety.

Sense making and decision making (September–November 2020): the second pandemic wave and staged lockdowns

In the aftermath of the summer months, Italians dealt with the second wave of COVID-19. The Campania and Lazio regions witnessed a spike in the number of people affected by the infectious disease. The number of cases doubled compared to what it was in March, at the infection's peak, and the country adopted new guidelines for containing the virus. All the regions were assigned labels based on risk: red=high risk, orange=medium risk, or yellow=low risk. Colors were assigned based on the level of disease reproduction in the region. The index of disease reproduction, R_0 (pronounced "R naught"), indicates how many people are infected by each sick person. The goal is to have an R_0 lower than 0, meaning that the number of people infected by a person who has COVID-19 is fewer than 0 and, therefore, the infection is phasing down. In this stage, new lockdowns were enacted – and some riots occurred. Notably, the parliament decided on October 7 to move the date of the end of the state of emergency forward to January 31, 2021, and people were again asked to wear masks outdoors. In this stage, regional governments were only given the authority by the central government to make the existing rules more stringent, not less. Meanwhile, new variants of the virus were discovered worldwide, and uncertainty ensued regarding whether the vaccine would be able to fight the novel coronavirus.

Terminating the crisis and learning from it (December 2020– January 2021): Phase 3 – the vaccine campaign

Toward the end of December 2020, the Italian government started its vaccination campaign, initially relying only on the Pfizer vaccine. All the vaccine doses were received in coordination with other European countries. The availability of vaccines then increased in January 2021, when the European Medicines Agency (EMA) approved the administration of the Moderna vaccine.

The Italian management of the COVID-19 crisis

Italy was the first Western country to deal with COVID-19. Its approach to the management of the COVID crisis resulted from several factors: the governance and leadership then in place, the unfolding of events as the crisis developed, and the socioeconomic context. As we examine the Italian structure of governance, we see issues of leadership, coordination, and collaboration in the multilevel governance contests that resulted from the interaction of municipal, regional, and central levels of the government. The sections that follow in this chapter describe the Italian system of governance, together with the issues of coordination and collaboration that arose in the COVID-19 crisis, providing a context for the governmental policies that were enacted to manage the crisis and its socioeconomic impacts.

The Italian structure of governance: coordination, collaboration, and leadership

At the national level, the Italian system of governance is altered in times of crisis. It is a consolidated praxis that the government responds to disasters such as earthquakes, waste management crises, and flooding with extraordinary powers. The state of emergency, a common procedure after natural disasters like earthquakes or floods, gives regional authorities special powers and cuts red tape. The department in charge of emergencies in Italy is the Department of Civil Protection. However, in responding to COVID-19, an important role was played by regional institutions. Even though catastrophes or disasters (such as earthquakes and floods) are usually managed at the national level, issues concerning health are under regional control. This is the result of the 2001 constitutional reform that gave more legislative powers to regions (Capano, 2020). Thus, Italy's 20 regions all had responsibilities for intervention on COVID-19, as a matter of public health.

The repercussions of a regionalist system were immense when it came to the implementation of mitigation measures, such as social distancing and the reduction of inter-regional mobility. A number of regulations were enacted at the national and the regional levels of government. During the initial response to COVID-19, 278 regulations were issued by the Italian government at the national level. Meanwhile, at the regional level, regulation was similarly prolific (Capano, 2020). In some cases, national and regional regulations conflicted with one another, causing confusion over what was really the appropriate action to take, or what was the right information to consider. This, in turn, negatively impacted the credibility of the message being conveyed to the public (Rodríguez et al., 2007), undermining the response efforts.

The relationship between the central government and regional powers was also undermined by the continuous delegitimization of powers. For example, the central government decided to adopt a country-wide reopening without consulting the regional governments, and then had to negotiate with those governments about how to proceed. On other occasions, the regional governments acted as though the central government did not deserve to be treated as legitimate. During the initial response, for instance, the governor of the Marche region closed all the schools without consulting the central government. And the governor of Sicily asked people coming from the northern part of Italy not to visit Sicily, despite a call from the national government to show internal solidarity (Ruiu, 2020).

In understanding the peculiar behavior of some regional governors, it is necessary to take into account the fact that governors' seats are determined through popular vote, while the central government seats are determined by the parliament. In June of 2020, the governor of the Campania region declared the region COVID free and invited people from other parts of Italy to take a vacation in Campania. In the following months, the Campania region, together with the Lazio region, had one of the highest rates of COVID-19 infection in the country. The declaration of the Campania as COVID free could have been inspired by political pressures on the part of the hospitality sector, since the governor (also known as the "sheriff") had previously been accused of being too harsh and causing unnecessary disruption to the tourism sector.

Governmental policies to manage the crisis and socioeconomic impacts

When COVID-19 struck, Italy was still recovering from the 2008 financial crisis, which heavily impacted its healthcare system.

Several social-distancing measures were implemented. These, and other containment measures, affected many aspects of social life, such as celebrating weddings and funerals and carrying out sporting events. During the lockdown phase, schools at all levels closed; distance learning was implemented from elementary school through university. With few exceptions, all commercial activities closed as well. The only shops that could remain open were pharmacies, grocery stores, and, later, shops for children's clothing. Transportation and mobility were reduced and, in some cases, banned entirely; there was a shutdown of air travel and all inter-regional movement ceased. Mobility was restricted to the point that people could only walk their pets in the immediate proximity of their own homes. All the containment measures were detailed in the decree, "I stay at home" ("Io resto a casa"), and lockdown and social distancing measures in Phase 1 were enforced through the imposition of fines. The Italian government levied fines ranging from 400 to 3,000 euros for failure to comply with social distancing and other measures aimed at containing the pandemic, such as wearing masks and not venturing out of the house. People who tested positive for

COVID-19 were charged with criminal penalties when they were found to have violated the quarantine. Punishments for these violations were even higher, consisting of incarceration, ranging from 3 to 18 months, and fines, ranging from 500 to 5,000 euros. Even more severe penalties were enforced for those who infected other people; in these cases, the penalty could be as high as a life sentence.

In the long run, it became clear that the lockdown policies were deepening disparities and previous vulnerabilities, as always happen in disasters. Thus, the national government issued three decrees aimed at mitigating the negative effects of the lockdown on the population: the "Cura Italia" decree, the "Rilancio" decree, and the "Agosto" decree, with the latter two decrees extending some of the measures detailed in the "Cura Italia" decree.

The "Cura Italia" decree touched upon several aspects of the lives of the Italian people, including labor, family, and support for businesses. At the same time, the national government encouraged both the private and public sectors to adopt smart working policies for their employees. Employees in the private sector were protected by policies preventing the possibility of being fired because of COVID-19. Also, financial support in the amount of 600 euros, later increased to 1,200 euros, was provided to workers to help them take care of their children. An additional 12 days of work permission was given to caregivers of impaired people.

People's behavior toward the changes induced by the crisis

The lockdown brought about a net rupture with people's daily established routines. In this kind of situation, people needed to reorganize their lives and make sense of their new reality. Even though everybody was equally exposed to the same containment measures, not everybody reacted in the same way, at least in Phase 1. People had to find a way of making sense out of this new reality – a reality in which the world as they knew it had suddenly changed.

The first perceptual change is related to people's sense of safety. COVID-19 is an invisible threat that changed people's daily lived realities. As we saw earlier, the government's duty to ensure their citizens' safety was accomplished by restricting people's freedom and mobility rights. These mitigation measures, implemented to slow down the diffusion of the virus, are related to changes in the sensescape, the economic sphere, and the social sphere. In the lockdown, confining people to their homes resulted in a difference in noise levels, due to decreased aerial, vehicular, and pedestrian traffic, along with an increase in sounds like ambulance sirens and church bells. Touch was also altered since, with confinement and social distancing in place, touching gestures typical of Italian culture (such as kissing, hugging, and shaking hands) abruptly stopped. Italians found themselves seeking new ways to rebuild their sense of community and different ways to spend their leisure time, in both the real and the virtual spaces. In their lockdown situation, particularly during Phase 1, Italians performed on their balconies, using them as an extension of their social space (Clinch, 2020) to socialize and build a renovated sense of community. Children drew rainbows with the slogan "everything will be fine" ("andra tutto bene") and hung them outside on the balcony. Social life moved into virtual life, with increased participation in online events, such as virtual fitness classes. From weekly mass to the customs surrounding Easter, religious observance became difficult, since gatherings were not allowed. The pandemic had a stark visual impact, as well. Both familiar local spaces and landmark places, such as St. Peter's Square at Easter, were left empty, while television was filled with dramatic images connected to the number of deaths each day.

Italians were mainly compliant with the lockdown rules, even though these rules were not always clear. While it is true that violation of the quarantine and other lockdown rules was punishable by law, Italians mainly complied because they understood the severity of the situation. However, there were government representatives and local authorities who sought to obtain their citizens' compliance through sarcasm, used as a means of social control. Consider, for example, the comment of Vincenzo De Luca, governor of the Campania region: "Where the f– are you all going? You and your dogs, which must have an inflamed prostate. We will send the police over with flamethrowers."

Going out to walk dogs was one of Italians' main excuses for leaving their homes during the lockdown. Gatherings to celebrate graduations, weddings, funerals, and masses constituted violations of the quarantine laws.

Problems arose with the beginning of Phase 2 and the country-wide staged reopening. In this phase, regional governments regained control of the pandemic management, and different regional leaders found themselves in competition with one another. This stage was characterized by mass confusion, due to the lack of clarity, coherence, and consistency in the protective behaviors and social practices being mandated. Initially, wearing a mask in public was not mandatory; then, when the situation was supposed to be improved, mask wearing in public became mandatory. There was uncertainty about how some businesses, especially in the entertainment sector, should operate, and how people's social practices should be modified to prevent a second wave of the virus.

This chapter has emphasized the decision makers' challenges in organizing the response and recovery effort as they sought to create support around governmental policies. It highlighted the situation of deep uncertainty in which both institutional decision makers and laypeople found themselves, as they were among the first in the world to respond to and recover from

COVID-19. This status of being first in responding to COVID-19 created a context of uncertainty concerning which protective actions to recommend and undertake, given the impossibility of knowing how long the crisis would remain a threat. Uncertainty also stemmed from the gradualness with which knowledge unfolded regarding how the virus spreads and how long it remains on surfaces. As the book moves on to consider Italy's COVID-19 response in greater detail, this chapter now concludes with an overview of the book's organization and structure.

Book overview

This book reports on data collected through an online survey that was administered on Italian social media, including Facebook and WhatsApp. The research was carried out in Italy from May 4 to June 3, 2020, during Phase 2 of the COVID-19 response and short-term recovery. The book organization revolves around the study's theoretical framework and methodology, along with the survey's macro-areas. The latter are related to the private/individual sphere, the public sphere, and reflections on the world post-COVID-19. The private sphere concerns people's cultural worldviews, risk perception, and the adoption of protective behaviors during the COVID-19 emergency, while the public sphere is structured around risk communication, people's support of public policies, and the trustworthiness of information sources and channels. Finally, the book reports on Italians' reflections about how they see their future and that of their country in the aftermath of COVID-19, concluding with suggestions for research, policy, and practice.

The book is critical because it focuses on the transition from Phase 1 to Phase 2 of the pandemic response, reporting on the short-term recovery that marked the country's staged reopening. The second phase of managing the COVID-19 pandemic was substantially different from the first phase (which had been oriented toward full-lockdown measures). It is important to remember that, during the first phase, the "Italy model" led the management of the COVID-19 crisis, at least in Europe. In contrast, the second phase arrived later for Italy than it did for the UK, Spain, Germany, and France.

The book also elaborates on, and then tests, a classic American risk communication and protective behavior model, Lindell and Perry's (2012) Protective Action Decision Making model, concerning COVID-19 in a foreign context. The study attempts to integrate this model by introducing ontological (in)security (Giddens, 1984) as an explanatory change factor. From a methodological standpoint, the study relies on a web survey. The use of online surveys to gather data in several thematic areas related to people's opinions and attitudes on the management of the public and private spheres can be considered a methodological innovation. Online surveys are usually short and concise instruments. This study has proven that people's engagement in online surveys can also exceed what we typically think about the demands of survey length and time required for completion. The reasons behind this increased willingness to invest time in the survey are likely found in the fact that the respondents' experience with the topic was pervasive and traumatic, coupled with the fact that the pandemic created a situation in which people were spending more time at home and online.

The remainder of the book is organized as follows. Chapter 2 introduces and discusses the social construction of reality, behavioral change, and risk communication, explaining the theoretical framework that drives the entire study. Chapter 3 reports on the methodological approach used. This chapter describes the research design and the sample, and it explains the reasons behind some of the methodological choices made in the study, such as the motivation for using principal component analysis and multiple correspondences. Chapters 4, 5, and 6 report on the findings of the study. Risk perception and protective behaviors are discussed in Chapter 4. The trustworthiness of sources and channels of information, together with their effects on ontological security, are covered in Chapter 5. And Chapter 6 offers a glimpse of how Italians thought of their future during Phase 2, as well as how they now judge the management of the pandemic. This final chapter also offers some concluding reflections on lessons to be learned, future research areas to be explored, and suggestions for policy and practice.

This book is comprised of four main, interdependent sections: (1) The background of COVID-19 management in Italy is given in Chapter 1; (2) the theoretical framework and methods of the study are given in Chapters 2 and 3; (3) the study findings are presented in Chapters 4 and 5; and (4) the study conclusions point the way toward life in a post-pandemic world in Chapter 6. The next chapter, Chapter 2, will lay out the theoretical framework of the study and will discuss the methods used. Thus, while Chapter 1 highlighted the deep state of uncertainty caused by the pandemic in both the Italian government and its citizenry, Chapter 2 will explain how the willingness to undertake certain protective behaviors and to support governmental regulations served as a way to restore or maintain a perceived certainty about the surrounding social and physical world.

References

Ansell, C., Boin, A., & Keller, A. (2010). Managing transboundary crises: Identifying the building blocks of an effective response system. *Journal of Contingencies and Crisis Management, 18*(4), 195–207.

Baekkeskov, E. (2015). Transboundary crises: Organization and coordination in pandemic influenza response. In R. Dahlberg, O. Rubin, & M. T. Vendelø (Eds.),

Disaster research: Multidisciplinary and international perspectives (pp. 189–206). Routledge.

Boin, A. (2019, January). The transboundary crisis: Why we are unprepared and the road ahead. *Journal of Contingencies and Crisis Management, 27*(1), 94–99.

Boin, A., & 't Hart, P. T. (2007). The crisis approach. In H. Rodríguez, W. Donner, & J. E. Trainor (Eds.), *Handbook of disaster research* (pp. 42–54). Springer.

Boin, A.,'t Hart, P., Stern, E., & Sundelius, B. (2005). *The politics of crisis management: Public leadership under pressure* (2nd ed.). Cambridge University Press.

Boin, A.,'t Hart, P., Stern, E., & Sundelius, B. (2017). *The politics of crisis management: Public leadership under pressure*. Cambridge University Press.

Capano, G. (2020). Policy design and state capacity in the COVID-19 emergency in Italy: If you are not prepared for the (un) expected, you can be only what you already are. *Policy and Society, 39*(3), 326–344.

Clinch, M. (2020, March 14). Italians are singing songs from their windows to boost morale during coronavirus lockdown. *CNBC*. www.cnbc.com/2020/03/14/coronavirus-lockdown-italians-are-singing-songs-from-balconies.html

Comfort, L. K. (2007). Crisis management in hindsight: Cognition, communication, coordination, and control. *Public Administration Review, 67*(s1), 189–197.

Comfort, L. K., Kapucu, N., Ko, K., Menoni, S., & Siciliano, M. (2020). Crisis decision-making on a global scale: Transition from cognition to collective action under threat of COVID-19. *Public Administration Review, 80*(4), 616–622.

Comitato di Esperti in Materia Economica e Sociale. (2020, June). *Iniziativa per il rilancio "Italia 2020–2022"* [Report for the president of the Council of Ministers]. https://img.ilgcdn.com/sites/default/files/documenti/1591638669-202006-RAPPORTO-FINALE-COMITATO-DI-ESPERTI-IN-MATERIA-ECONOMICA-E-SOCIALE.pdf

Decreto del Presidente del Consiglio dei Ministri. (2020, March 9). *Ulteriori disposizioni attuative del decreto-legge 23 febbraio 2020, n. 6, recante misure urgenti in materia di contenimento e gestione dell'emergenza epidemiologica da COVID-19, applicabili sull'intero territorio nazionale.* (GU Serie Generale n.64 del 11/03/2020). www.gazzettaufficiale.it/eli/id/2020/03/11/20A01605/sg

Decreto del Presidente del Consiglio dei Ministri. (2020, April 26). *Ulteriori disposizioni attuative del decreto-legge 23 febbraio 2020, n. 6, recante misure urgenti in materia di contenimento e gestione dell'emergenza epidemiologica da COVID-19, applicabili sull'intero territorio nazionale.* (GU Serie Generale n.108 del 27/04/2020). www.gazzettaufficiale.it/eli/id/2020/04/27/20A02352/sg

Giddens, A. (1984). *The constitution of society: Outline of the theory of structuration.* University of California Press.

Goldin, I., & Mariathasan, M. (2014). *The butterfly defect: How globalization creates systemic risks, and what to do about it.* Princeton University Press.

Il Messagero (2020, March 8). *Coronavirus, a Milano la fuga dalla zona rossa: Folla alla stazione di Porta Garibaldi.* www.ilmessaggero.it/italia/coronavirus_milano_fuga_milano-trena-romani-5097472.html

Jervis, R. (2017). *Perception and misperception in international politics* (New ed.) Princeton University Press. (Original work published 1976)

Johns Hopkins University & Medicine (2021, January 8). *COVID-19 dashboard by the Center for Systems Science and Engineering (CSSE) at Johns Hopkins*

University (JHU). Coronavirus Resource Center, Johns Hopkins University & Medicine. https://coronavirus.jhu.edu/ map.html

La Repubblica. (2020, March 1). Decreto coronavirus, ecco tutte le misure del governo: L'atto firmato dal Presidente del Consiglio: l'Italia divisa in tre zone. *"Pieno sostegno" dell'Omms*. www.repubblica.it/politica/2020/03/01/news/coronavirus_misure_governo-249980561/

Lebow, R. N., & Stein, J. G. (1994). *We all lost the cold war* (Vol. 55). Princeton University Press.

Leonard, J. (2020, March 23). What we know about the plague of ancient Athens. *Greece Is*. www.greece-is.com/what-we-know- about-the-plague-of-ancient-athens

Lindell, M. K., & Perry, R. W. (2012). The protective action decision model: Theoretical modifications and additional evidence. *Risk Analysis: An International Journal, 32*(4), 616–632.

Macrae, F. (2010, February 17). The "false" pandemic: Drug firms cashed in on scare over swine flu, claims Euro health chief. *The Daily Mail*. www.dailymail.co.uk/news/article-1242147/The-false-pandemic-Drug-firms-cashed-scare-swine-flu-claims-Euro-health-chief.html

Marantette, K. (2021, February 19). What the COVID-19 pandemic looks like compared to the 1918 influenza pandemic. *WLNS 6 News*. www.wlns.com/news/what-the-covid-19-pandemic-looks-like-compared-to-the-1918-influenza-pandemic/

Mineo, L. (2020, May 19). "The lesson is to never forget": Harvard expert compares 1918 flu, COVID-19. *Harvard Gazette*. www.new.harvard.edu/gazette/story/2020/05/harvard-expert-compares-1918-flu-covid-19/

Nacoti, M., Ciocca, A., Giupponi, A., Brambillasca, P., Lussana, F., Pisano, M., & Longhi, L. (2020). At the epicenter of the COVID-19 pandemic and humanitarian crises in Italy: Changing perspectives on preparation and mitigation. *NEJM Catalyst Innovations in Care Delivery, 1*(2). https://catalyst.nejm.org/doi/full/10.1056/CAT.20.0080

O'Dowd, A. (2010, June 7). Council of Europe condemns "unjustified" scare over swine flu. *The BMJ*. https://bmj.com/content/340/bmj.c3033. https://doi.org/10.1136/bmj.c3033

Online Etymology Dictionary. (n.d.). *Pandemic*. www.etymonline.com/ search?q= pandemic

Parker, C. F., & Stern, E. K. (2002). Blindsided? September 11th and the origins of strategic surprise. *Political Psychology, 23*(3), 601–630.

Perrow, C. (2011). *Normal accidents: Living with high-risk technologies* (Updated ed.). Princeton University Press. (Original work published 1984).

Quarantelli, E. L. (2000). *Emergencies, disasters, and catastrophes are different phenomena* [Preliminary paper #304]. University of Delaware Disaster Research Center.

Quarantelli, E. L. (2006). *The disasters of the 21st century: A mixture of new, old, and mixed types* [Preliminary paper #353]. University of Delaware Disaster Research Center.

Quarantelli, E. L., Lagadec, P., & Boin, A. (2006). A heuristic approach to future disasters and crises: New, old and in-between types. In H. Rodriguez, E. L.

Quarantelli, & R. R. Dynes (Eds.), *Handbook of disaster research* (pp. 16–41). Springer.

Rodríguez, H., Díaz, W., Santos, J., & Aguirre, B. (2007). Communicating risk and uncertainty: Science, technology, and disasters at the crossroads. In H. Rodríguez, E. Quarantelli, & R. Dynes (Eds.), *Handbook of disaster research* (pp. 476–488). Springer.

Rosenthal, U. (1998). Future disasters, future definitions. In E. Quarantelli (Ed.), *What is a disaster? Perspectives on the question* (pp. 146–160). Routledge.

Ruiu, M. L. (2020). Mismanagement of Covid-19: Lessons learned from Italy. *Journal of Risk Research*, *23*(7–8), 1007–1020.

Stern, E. (2013). Preparing: The sixth task of crisis leadership. *Journal of Leadership Studies*, *7*(3), 51–56.

't Hart, P., & Boin, A. (2001). Between crisis and normalcy: The long shadow of post-crisis politics. In U. Rosenthal, A. Boin, & L. K. Comfort (Eds.), *Managing crises: Threats, dilemmas, opportunities* (pp. 28–46). Charles C. Thomas.

World Health Organization. (2009). *Current WHO phase of pandemic alert for Pandemic (H1N1) 2009*. Retrieved February 10, 2021 from www.who.int/csr/disease/swineflu/phase/en

World Health Organization. (2010). *What is a pandemic?* www.who.int/csr/disease/swineflu/frequently_asked_questions/pandemic/en/

World Health Organization. (2020, March 11). *WHO director-general's opening remarks at the media briefing on COVID-19*. World Health Organization. www.who.int/director-general/speeches/detail/who-director-general-s-opening-remarks-at-the-media-briefing-on-covid-19–11-march-2020

Yuko, E. (2020, September 12). 13 ways coronavirus is different from all epidemics through history. *Readers Digest*. www.rd.com/ways-coronavirus-is-different-from-other-epidemics

2 Risk communication and the social construction of COVID-19

Introduction

This chapter discusses the theoretical background of the study. The study assumes that, beyond the epidemiological aspects of COVID-19, there is also a social construction dimension (Berger & Luckman, 1966) of the pandemic to consider. Whatever else it may be, the COVID-19 pandemic is a socially constructed event. This means that it is important to recognize the relationship between the micro (individual) level and the macro (societal) level in this crisis, along with the fact that these two levels influence each other in determining distinctive behavioral patterns (actions), such as the preference for one policy over another and the willingness to adopt a specific protective behavior.

Risk communication is part of the meaning making of every governmental crisis (Ansell et al., 2010), and it is a crucial component of any modern emergency management system. After making sense of the crisis, governments need to update the public on the situation and indicate protective actions. Thus, depending on the kind of threat to which a community is exposed, governments issue warnings. Warnings provide information about the danger and the protective actions to be undertaken. Most emergency risk communication is predicated on the assumption that the warning system's technical and social improvements can motivate people to take protective actions. Contrary to popular belief, though, communicating risks to the population and modifying their behavior is not an automatic result of issuing a warning. Making significant improvements to the warning system is difficult since the relationship between issuing warnings and undertaking protective behavior is complex and multifactorial (e.g., Floyd et al., 2000; Gladwin, 2007; Lindell & Perry, 1997, 2004, 2012; Mileti & Sorensen, 1990).

The importance of understanding what motivates people's protective behaviors and their support toward governmental regulations becomes particularly important in a context like that generated by COVID-19, where

DOI: 10.4324/9781003187752-2

protective behaviors need to be sustained for an unknown amount of time and governmental regulations for containing the disease disrupt people's routines. The study assumes that different profiles/groups of people adhere to different constructions of social reality. Depending on how reality is socially constructed, people will experience more or less intense feelings of imbalance regarding the previous way of living their lives and making sense of themselves and the social and physical reality surrounding them. People will try to restore the feeling of security by adopting or not adopting new behaviors and attitudes. In essence, this chapter explains what motivates people's adoption or support of new behaviors and policies.

We will take a look at the concepts of the social construction of reality (Berger & Luckman, 1966) and risk communication, with a view toward protective action models in disaster science. The chapter will then offer a theoretical framework aimed at integrating a particular protective action model, the Protective Action Decision Making (PADM) model (Lindell & Perry, 2012), with the concept of ontological (in)security. Finally, the study's research questions will be put forward.

The social construction of reality and behavioral change

When thinking of reality as socially constructed, the first questions one might have relate to the process (i.e., how this happens) and the outcome (i.e., how a socially constructed reality is kept in place, and how it is eventually replaced by another one).

The process through which reality is socially constructed is relatively straightforward in theory, but it is very complex in practice. To understand how the process works, we need to answer the following questions: How do we make sense of our physical and social environment? How do we know how to behave in different contexts?

Society is understood through interpretative schemes. How to interpret society, what the society's norms are, and how to behave in this society are all things we learn. For instance, we know the meaning of safety and who is responsible for it through primary and secondary socialization. Primary socialization happens during the early stage of a child's development, through the family. The process of secondary socialization occurs outside the home, through television, peers, neighbors, and school. Socialization processes provide individuals with the tools to reassess their reality and to decide what the best course of action is in unclear situations.

In essence, a socially constructed reality (Berger & Luckman, 1966) is based on a shared set of values and beliefs reproduced through actions or procedures. These actions and procedures are not meaningless but carry implicit meanings that can be universal and recognizable by society at large;

or they can be more particular and recognizable by members of a specific community. Meanings and actions reproduce each other in a recursive way (Giddens, 1984). That is to say, the interaction between the macro and the micro levels determines people's routine actions and societal rules. Individuals tend to reproduce and recognize behavioral patterns and norms. This mechanism is explained by Giddens's (1984) *structuration theory*, which highlights how reality is formed by the interaction of structure and action. Reality is socially constructed through the daily production and reproduction of actions, norms, and procedures. A way for people to reproduce reality is through mechanisms of accountability. People reproduce a shared understanding of the world by being accountable to one another. Sometimes, some of the reproduced processes are so ingrained in people's lives that they reproduce them effortlessly; these individuals do not even have to think about the actions that they are going to take. In other cases, this process is more complex. These are the cases when the best response is not clear cut and the individual needs to actually think about how to behave.

Human beings run their lives based on well-established conventions that make performance of daily tasks easy and intuitive. For instance, people know how to behave at the grocery store: entering, selecting the items they need, and then paying. Some experiences or contexts can be new to people and intimidating because they leave them unsure of how to behave. The uncertainty regarding how to act can result from interacting in an unfamiliar environment. Familiarity with contexts, situations, rules, and procedures is a result of routinization. *Routinization* is "the habitual taken for granted character of the vast bulk of the activities of the day to day social life; the prevalence of familiar styles and forms of conduct both supporting and supported by a sense of ontological security" (Giddens, 1984, p. 376). When addressing the routinization of daily life, the concept of ontological security is crucial, since it supports routines and is simultaneously supported by the routines as well.

Ontological security is "the confidence or trust that the natural or social world is as it appears to be, including the basic existential parameters of self and social identity" (Giddens, 1984, p. 375). The concept of ontological security highlights how people make sense of the surrounding world and of themselves in a shared way. In essence, people have confidence that interpreting reality and making sense of themselves is the same for other people. Routine behaviors and the predictability of the physical and social environment provide people with a sense of ontological security.

However, if people tend to learn and reproduce how to behave, how does behavioral *change* happen? What motivates change? To answer this question, we need to understand how ontological (in)security works. If routines are supported by ontological security, behavioral change, in answer to uncertainty or routine disruption, might result from a person's experience of different degrees of *ontological insecurity*.

Different degrees of ontological security can be experienced in relation to the elements that make it up. The core element of ontological (in)security is trust/distrust or confidence/uncertainty, triggered by (1) certainty/uncertainty about the surrounding physical world; (2) certainty/uncertainty about the surrounding social world; and (3) confidence in one's identity (Velotti & Justice, 2016, p. 9). The concept of trust/confidence refers to people's attempt to make sense of the surrounding physical and social world in a shared way. Trust, or its absence, is triggered by uncertainty in the shared understanding of both the social and physical environment and one's identity. Identity is formed and reinforced through interaction with people. Individual characteristics and group identities make up one's identity. This identity results from belonging to several groups and subgroups. These groups, in turn, are identified by gender, family, peers, professional groups, socioeconomic status, cultural worldviews, and religious and political orientation, among other things. Thus, one's identity is the result of a complex set of values and expectations. For instance, some studies report that people's professional and personal identities coincide (Roberts et al., 2020; Velotti, 2016; Velotti & Justice, 2016). Cultural worldviews are also an important component of people's identity. In understanding the elements that trigger different degrees of ontological (in)security, *cultural worldview*, as a strong component of a person's identity, deserves its own discussion.

Cultural worldviews, or individuals' cultural biases, are "culturally constrained rationalities" (Song et al., 2014, p. 530) that may determine whether a specific worldview influences one's support for governmental regulation or adoption of protective behaviors. Cultural theory (Douglas & Wildavsky, 1982a, 1982b; Schwarz & Thompson, 1990; Thompson, 2008; Thompson et al., 1990; Wildavsky, 1987) identifies four kinds of cultural worldviews:

1. Individualism;
2. Egalitarianism;
3. Hierarchism; and
4. Fatalism.

Individualists "dislike authority, external prescription, and the idea of equity based on equal outcomes rather than individual merit and effort" (Song et al., 2014, p. 532). In essence, individualists perceive the relationships with other people as transactional. The *egalitarians* believe that the community (and not the experts) should rule, since the community is less concerned with social deviance. Egalitarians also believe in the role of government. The *hierarchic* orientation makes people more prone to believe that a society will benefit from specialized jobs and experts. Hierarchism also does not like social deviance, because hierarchists firmly believe in a "superior/subordinate form of social relationships" (Douglas & Wildavsky, 1982a). Unlike hierarchism, *fatalists* believe in fate and luck, and they tend not to engage in social relationships. The literature on public policy suggests that people with a less political orientation will be less affected by cultural worldviews.

Johnson and Swedlow (2021) state that cultural theory contributes to understanding how cultural worldviews work in relation to one's identity. These authors explain that cultural worldviews, as part of people's group identities, produce at least two effects on identity: (1) Identity protective cultural cognition (Cohen, 2003; Finucane et al., 2000; Kahan et al., 2007) and (2) cultural identity affirmation (Cohen, 2000; Cohen et al., 2007; Cultural Cognition Project, 2007). According to *identity protective cultural cognition*, people tend to behave in a way that is coherent with their identity. *Cultural identity affirmation* occurs when people have to deal with facts, situations, and events that could potentially threaten their cultural worldviews. People will tend to support more the information that is presented as supportive of their own cultural worldview.

In a nutshell, reality is socially constructed through mechanisms of accountability with which people reproduce societal and group expectations. Behavioral change might occur in response to a changed social and physical environment; and when one's identity is threatened, this generates different degrees of ontological insecurity. An important takeaway of understanding reality as socially constructed, according to cultural theory, is that "risks are socially selected and at least in part socially constructed to serve the social relations of those perceiving and analyzing them" (Johnson & Swedlow, 2021, p. 430). Being able to better comprehend the social construction of risk is important for improving systems of risk communication.

Risk communication and protective action models in disaster science

Risk communication is a crucial component of any modern emergency management system. However, contrary to popular belief, communicating risks to the population and modifying their behavior is not an automatic result of issuing a warning. Risk communication is a multifactorial and multistage process (Floyd et al., 2000; Gladwin, 2007; Lindell & Perry, 1997, 2004, 2012; Mileti & Sorensen, 1990).

The warning process has been described in the warning literature in several ways, according to different perspectives, but mainly all converge toward the same factors. In disaster science, the Protective Action Decision

Making model (Lindell & Perry, 1997, 2004, 2012) and the Mileti and Sorensen model (1990) are the two main models describing how people respond to warnings (Donner et al., 2012; Nagele & Trainor, 2012). Sociologists (Sorensen, 2000) refer to the warning process as both a sequential process (Mileti & Sorensen, 1990) and a stage process (Lindell & Perry, 1997, 2004). In this next section of the chapter, we will explore these two models.

Mileti and Sorenson's model (1990) states that an effective warning system is made up of three integrated subsystems. The subsystems taken into account are:

1. *The detection subsystem*, in which organizations in charge of detecting hazards scan and assess the environment. For instance, in case of a pandemic, the World Health Organization (WHO) is in charge of raising the status of an infectious disease to pandemic level and then issuing a worldwide warning.
2. *The management subsystem* involves how institutions interpret and use data. For instance, how do the local and national levels of government make sense of, and act on, the information about an incumbent hazard, such as a pandemic?
3. *The response subsystem*, whereby people receive the warning and respond to it. The response subsystem is, in turn, made up of two constituent parts: interpretation and response. The *interpretation* consists of the following components: hearing the warning; understanding the warning; believing, personalizing, and confirming the warning; deciding if the warning pertains to the recipient; determining if protective action is necessary and feasible; and determining which action to take. Hearing and understanding the warnings are the enabling parts of the response subsystem. Individuals need to be exposed to a warning if they are going to act on it and, more importantly, they need to understand what the warning is about and what protective behavior needs to be undertaken in response. The second component is the actual *response* itself. This may involve undertaking a protective behavior, such as evacuating an area, moving furniture indoors, wearing a mask, and avoiding gatherings, to mention a few possible responses.

The second model describing how people interpret and respond to warnings is the Lindell and Perry model (1997, 2012). Their *Protective Action Decision Making (PADM)* model is aimed, together with situational facilitators and impediments, at predicting how behavioral responses are produced in risk situations. The PADM model (Lindell & Perry, 2012) is used to understand people's behavior during several types of hazards, including health crises such as the H1N1 pandemic influenza in China, the 2013 Chinese

H7N9 influenza outbreak (Wang et al., 2018), the COVID-19 crisis, and people's perceptions and expected responses to influenza (Wei et al., 2020). The majority of these studies were carried out in China and the USA. The PADM model delineates the following five stages:

1. Processing information derived from social and environmental cues with messages that social sources transmit through communication channels to those at risk;
2. Pre-decisional processes made up of exposure, attention, and comprehension of the warning;
3. Core perceptions, such as threat perceptions, protective action perceptions, and stakeholder perceptions;
4. Protective action decision making; and
5. *Behavioral response*, which refers to three main behaviors: information search, protective response, and emotion-focused coping.

Stage one of the PADM model presupposes that the activation of pre-decisional processes, in stage two, can be determined by the interpretation of environmental and social cues and/or official warning messages, sources and channels of information, and the trustworthiness of the stakeholders, together with receiver characteristics. Social cues are signals that people receive by observing other people's behaviors. Consider the people who might be tempted not to shelter in place when given a tornado warning, because the people surrounding them are not undertaking any protective behavior. People get environmental cues through their senses: smell, sight, and hearing. For instance, people make sense of a tornado's severity by seeing it or hearing it; many people have reported hearing a sound similar to that of a freight train as a tornado approaches. People make sense of earthquakes because of the tremblor. A limited number of studies use the PADM to consider social and environmental cues in the context of a pandemic or infectious disease. And this is undoubtedly because the virus is not seen and not felt, but can only be perceived as emotionally and geographically near or distant in its manifestations (number of infected persons, beds in hospitals saturated, intensive care units clogged, and so on) – images that result in a direct and indirect experience of the virus, its consequences, and its legacies on various systems surrounding the individual.

The PADM model also considers the sources, the channels of information, and the trustworthiness of the stakeholders. Sources of information may be local or national authorities, relatives, friends, etc. "Channels" here refers to television stations, newspapers, radio broadcasts, and social media. Last, but not least, an important role is assumed by the characteristics of the warning message itself. A well-designed warning should suggest more than one protective action (Lindell & Perry, 2012), and it should be both credible and consistent. The more a message is released by a credible source or by a source that dictates compliance with it, the more people will comply, and there will be a reduction in warning ambiguity (Lindell & Perry, 2012). Consider, by way of example, the enforcement of fines for not complying with a behavior.

The receiver characteristics are also part of people's decision-making process since they constitute the starting point for determining opportunities or constraints toward adopting recommended protective actions. Examples of receiver characteristics are "physical (e.g., strength), psychomotor (e.g., vision and hearing), and cognitive (e.g., primary and secondary languages as well as their mental models/schemas) abilities as well as their economic (money and vehicles) and social (friends, relatives, neighbors, and co-workers) resources" (Lindell & Perry, 2012, p. 617).

Social and environmental context variables, together with characteristics of the warning message, types of sources and channels, stakeholders' trustworthiness, and receiver characteristics, impact stage two of the PADM model, which consists of three pre-decisional processes: (1) people's exposure, (2) their attention, and (3) their comprehension of the message. For people to undertake any protective action, a message needs to reach them. For example, if a televised warning is going to reach someone, that person needs to have the television on and be watching it during the broadcast warning. And yet, being exposed to a warning is not enough; people also need to pay attention to the warning in order to start processing the information conveyed in it.

Once a person is exposed and gives attention to the warning message, the next step is to comprehend it. People might not understand the message's content, because of impairments (such as having hearing or sight problems), language barriers, or excessive jargon used to deliver the message. The pre-decisional processes of exposure, attention, and comprehension happen intuitively. At the same time, the perception of the threat, the stakeholders, and the protective action, which are part of stage three of the model, can either be automatic or the result of some reflection. The PADM model relies on people's mental models in perceiving risk and undertaking protective action. The mental model approach can help us to understand how people create a schema in their minds about how a disaster will unfold and what measures are appropriate to take.

Risk perception is based on perceived consequences, probabilities, dread (fear), and unknown risks (Slovic et al., 1980). It is also related to hazard experience and proximity. Hazard experience has to do with the consequences that an individual will face if exposed to the threat. For instance, will they die? The model also states that panic rarely occurs. The willingness of people to take hold of the warnings issued by stakeholders (such as authorities and experts) depends on the perceived trustworthiness of these individuals and entities. Before deciding whether to take protective action, and what kind of action to undertake, people go through the risk identification stage to evaluate if a real threat exists; then they proceed to the risk assessment stage to decide if it is necessary to take some kind of action. And any decision an individual makes, which is part of the final stage of the model, will also have to take into consideration situational facilitators and impediments to that decision. In the case of a hurricane, for example, people might want to evacuate, but they may not have a car to leave the area at risk – as was the case during Hurricane Katrina.

The PADM and other risk communication models aim to describe how people process information, take protective actions, and/or adopt protective behaviors. While all these models represent stages through which people change their behaviors, none of them pinpoint what *motivates* change and, in particular, "what motivates people to undertake protective actions" (Lindell & Perry, 2012, p. 625). In understanding people's behaviors during a pandemic or infectious disease crisis, it is essential to consider that the kind of protective behavior under consideration is not a temporary behavioral adjustment; it can be long-lasting. Another aspect to consider is the onset of the hazard. A pandemic is an unfolding event that needs to be recognized as a crisis. In COVID-19, people went through the stages of the PADM model more slowly than they would have if they had to decide on protective actions to be undertaken during an event like a tornado. In addition, one of the things that the PADM model does not address is the level of uncertainty elicited by some hazards. It is worth noting that a pandemic as a hazard is better understood in some specific regions of the world than in others. Italy, for example, has dealt only marginally with pandemics.

The social construction of COVID-19, risk communication, and behavioral change

This study expands the PADM model and explores its antecedents in the context of the COVID-19 pandemic health crisis. One of the main assumptions of PADM is that risk perception drives the adoption of protective behaviors. In this study, we expand the PADM model by adding the concept of ontological (in)security as an element that motivates people to protect themselves and others. The idea of ontological security helps explore the effects of the social construction of individuals' reality. Ontological security can be a good starting point to understand people's motivation to change behavior or attitude and sustain it through time. In COVID-19, adopting some protective behaviors is less temporary than what is required to respond to other hazards. For instance, the lockdown requires a change in behavior that is not short-lived but rather needs to be sustained through time. Wearing a mask, practicing social distancing, and changing ways of employing leisure time are all behavioral changes that need to be sustained, at least until herd immunity is reached.

The PADM (Lindell & Perry, 1997, 2012) and the Mileti and Sorensen (1990) models of risk communication both highlight the dimension of risk perception as driving people's behavioral change. However, risk, as proposed by the theory of the culture, can also be socially constructed. The concept of ontological (in)security (Giddens, 1984) offers a good synthesis of change resulting from both perceived and socially constructed risk. The core concept of ontological (in)security is trust. Trust is a concept that has been used extensively in disaster science, though it is not fully operationalized (Liu & Mehta, 2020). Velotti and Justice (2016) operationalized the concept of ontological security as being made up of trust/confidence that the perception of the physical surrounding and social environment, along with one's identity, are shared. Thus, it is possible to highlight three dimensions of trust/confidence, which relate to the physical environment, the social environment, and the personal sphere. These dimensions of trust have here additionally been operationalized by integrating them with similar concepts pertaining to the PADM model (see Figure 2.1).

In the PADM model, "trust" is primarily used to indicate trust in sources and channels, and in stakeholders' trustworthiness. This study expands the PADM model's use of "trust" and defines "trust" as triggered by certainty about the physical and social world, as well as the ability to retain one's identity and control in their personal sphere. The concepts of physical and social certainty cover those of environmental and social cues in the PADM

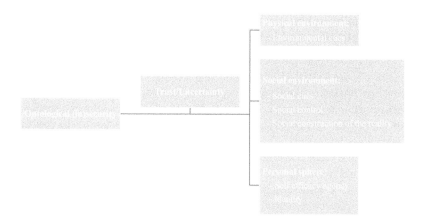

Figure 2.1 Core elements of ontological (in)security

model. However, we expand upon the concept of social cues by also adding the entire meaning-making process during a hazardous event.

The concept of environmental cues, or certainty about the surrounding environment during an infectious disease, is a challenging one to conceptualize, since no signs/cues can be seen, smelled, heard, or tasted. Thus, the disease gets recognized through its symptoms or images, and through sounds, tastes, or odors that are associated with it. The association of images, sounds, tastes, and odors (or the lack thereof) is a less direct experience because it is socially constructed. By way of example, it is customary to toll the church bells in Italy when someone dies. But during the peak of COVID-19, the bells tolled so much that some villages stopped the tradition to avoid affecting people further. During the pandemic, a new soundscape served as a reminder of a changed physical and social world. As another example, people noticed a reduction in aircraft and vehicular traffic volume (Spennemann & Parker, 2020). Yet, there was also an increase in the sound of ambulances and temporary noise generated by hand-clapping and singing from balconies as a sign of appreciation for nurses and doctors. People can also get environmental exposure to the disease through their senses, by either direct or indirect experience. With COVID-19, people may have caught the disease or have been exposed to the infection through their friends or through second-hand experience.

Certainty about the social world results from other people's compliance with the recommended protective actions and the new meanings that need to be created to adapt to the new reality. A socially constructed dimension that relates to trust is safety. During the pandemic, safety is dependent both on the government's ability to solve the crisis and on individual and collective responsibility. Everyone is responsible for other people's safety, as well as for their own. Trust or confidence manifests in different ways through people's feelings of self-efficacy – that is, through an individual's confidence about contributing to other people's safety through the adoption of the containment measures suggested by the government.

The mechanisms for ensuring that new norms and actions get reproduced can be both legal and social. To ensure that people wear masks or respect the social distance policy, people might use social control, like reprimanding those who are not conforming. The use of social control is a way for people to make sure that the surrounding world is under control and approaching the safety standards shared by the government with the community. People may or may not have confidence that governments at different levels (national, regional, and local) will solve an eventual second wave of COVID-19. One's responsibility toward achieving collective safety is so ingrained that a noncompliant behavior with the containment measures is sanctioned through the issuing of fines.

The response effectiveness to COVID-19 results from collective behavior in which everybody protects others by protecting themselves. A pandemic, as an infectious disease, requires people to take responsibility for their safety and for that of others. Thus, the dimension of personal responsibility for other people's safety and wellbeing becomes a prominent variable to be examined. Individuals might decide that they do not need to engage in protective behaviors, but they will do so because they believe their behavior affects the disease's containment. Responsibility for oneself and others also needs to be understood in the context of people's perceived self-efficacy – that is, in the context of the belief that one's action will be able to impact the outcome positively.

In order to be accountable to each other through their protective actions, people need to collectively reframe how they understand the physical and social world, balancing it against their identities. The attempts to retain one's identity by discarding any identity-threatening piece of information or event (Cohen, 2000; Cohen et al., 2007; Cultural Cognition Project, 2007), as well as cultural cognitions (Cohen, 2003; Finucane et al., 2000; Kahan et al., 2007), need to be taken into account. For instance, social distancing rules and lockdowns challenge how we envision the hospitality business. Restaurants, hotels, and many tourist locations – along with all the people reliant on those sectors – need to rethink how they operate. Restaurants need to respect the three feet of distance between tables; restaurant owners cannot approach clients anymore with a handshake. Most of the gestures of conviviality are lost. New ways of running a business might impact the business owners' identities, generating insecurity as new rules are established. There is uncertainty over whether there will be acceptance of the new modus operandi by customers. In other cases, young caregivers might feel that they are personally not at risk, but that people in their care might be; this might impact their identity as good caregivers and motivate them to undertake protective actions. Young people with an egalitarian worldview might comply with recommended protective actions because of their desire to contribute to their community's wellness. Yet, the lack of compliance with protective measures, such as the wearing of masks and gloves, and the lack of compliance with social distancing norms, might be a way to deny the existence of COVID-19 as a threat, in both the individualist and the fatalist cultural worldviews. This attitude, usually labeled irrational, is based on people's attempts to retain their professional identity. Consider the case of people working for the hospitality sector, such as restaurateurs and coffee shop owners. Their professional identity is made up of social practices, such as shaking hands, but this is upended by the new reality dictated by the pandemic. And yet, this does not mean that all the people working in the hospitality sector will have a defiant behavior toward adopting protective actions.

Ontological (in)security must be considered on a continuum, as occurring in different degrees (low to high). Not everyone will experience the same degree of ontological (in)security, but they will tend to perceive different gradations of it in accordance with a specific identity construction. Thus, based on the theoretical framework, the following research questions have been identified:

* Does ontological (in)security affect one's undertaking of protective behavior/s?

   * Is the undertaking of protective behavior/s affected by people's individual characteristics, such as gender, socioeconomic status, age, level of education, religious belief, profession, political orientation, and cultural worldview?
   * Is the undertaking of protective behavior/s affected by people's direct and indirect experience and their perception of risk or social and environmental cues?
   * Is the preference for a specific protective behavior (or set of protective measures) affected by people's understanding of the phenomenon, type of usage, and conferral of trust to different channels and sources of information, and confidence in the government and healthcare system?
   * What are the factors determining various degrees of ontological (in)security?

Starting from an empirical survey conducted during the second phase of the first lockdown in Italy, the fourth chapter of this volume reports on the influences of the individual sphere on the decision-making mechanism for the adoption of protective behaviors, and on respondents' support toward governmental regulations. In particular, Chapter 4 will look at the undertaking of protective behaviors on the basis of the three core elements that trigger trust: (1) Trust in the social environment, (2) trust in the physical environment, and (3) trust in the ability to reproduce one's identity and self-efficacy, which relates to people's personal sphere. The fifth chapter will focus on the influences of the wider public and the social sphere. To the main question that guides the study, we will try to find an integrated answer in the sixth and last chapter, which tries to bring to the system all the integrations made to the PADM model that have been coupled with reflections on the ontological (in)security factor and the social construction of COVID-19. But before we move into that analysis, we must first turn our attention to the methodology supporting the entire study, in Chapter 3. This chapter will address issues related to the strategy of sampling, the structure of the questionnaire, and the strategy of analysis.

References

Ansell, C., Boin, A., & Keller, A. (2010). Managing transboundary crises: Identifying the building blocks of an effective response system. *Journal of Contingencies and Crisis Management, 18*(4), 195–207.

Berger, P. L., & Luckman, T. (1966). *The social construction of reality: A treatise in the sociology of knowledge.* Anchor.

Cohen, G. L. (2003). Party over policy: The dominating impact of group influence on political beliefs. *Journal of Personal Social Psychology, 85*(5), 808–822.

Cohen, G. L., Sherman, D. K., Bastardi, A., Hsu, L., McGoey, M., & Ross, L. (2007). Bridging the partisan divide: Self-affirmation reduces ideological closed-mindedness and inflexibility in negotiation. *Journal of Personality and Social Psychology, 93*(3), 415.

Cohen, M. P. (2000). *Risk, vulnerability, and disaster prevention in large cities.* [Working paper.] Lincoln Institute of Land Policy. www.alnap.org/help-library/risk-vulnerability-and-disaster-prevention-in-large-cities

Cultural Cognition Project. (2007). *The second national risk and culture study: Making sense of – and making progress in – the American culture war of fact.* Yale Law School. www.culturalcognition.net/projects/second-national-risk-culture-study.html

Donner, W. R., Rodriguez, H., & Diaz, W. (2012). Tornado warnings in three southern states: A qualitative analysis of public response patterns. *Journal of Homeland Security and Emergency Management, 9*(2). https://doi.org/10.1515/1547-7355.1955

Douglas, M., & Wildavsky, A. (1982a). *Risk and culture: An essay on the selection of technical and environmental dangers.* Berkeley, CA: University of California Press.

Douglas, M., & Wildavsky, A. (1982b). How can we know the risks we face? Why risk selection is a social process. *Risk Analysis, 2*(2), 49–51

Finucane, M. L., Slovic, P., Mertz, C. K., Flynn, J., & Satterfield, T. A. (2000). Gender, race, and perceived risk: The white male effect. *Health, Risk & Society, 2*(2), 159–172.

Floyd, D. L., Prentice-Dunn, S., & Rogers, R. W. (2000). A meta-analysis of research on protection motivation theory. *Journal of Applied Social Psychology, 30*(2), 407–429.

Giddens, A. (1984). *The constitution of society: Outline of the theory of structuration.* University of California Press.

Gladwin, H. (2007). Evacuation decision making and behavioral responses: Individual and household. *Natural Hazards Review, 8*(3), 69–77.

Johnson, B. B., & Swedlow, B. (2021). Cultural theory's contributions to risk analysis: A thematic review with directions and resources for further research. *Risk Analysis, 41*(3), 429–455.

Kahan, D. M., Braman, D., Gastil, J., Slovic, P., & Mertz, C. K. (2007). Culture and identity-protective cognition: Explaining the white-male effect in risk perception. *Journal of Empirical Legal Studies, 4*(3), 465–505.

Lindell, M. K., & Perry, R. W. (1997). Principles for managing community relocation as a hazard mitigation measure. *Contingencies and Crisis Management, 5*(1), 49–59.

Lindell, M. K., & Perry, R. W. (2004). Approaches to influencing hazard adjustment adoption. In M. K. Lindell & R. W. Perry (Eds.), *Communicating environmental risk in multiethnic communities* (pp. 171–215). Sage.

Lindell, M. K., & Perry, R. W. (2012). The protective action decision model: Theoretical modifications and additional evidence. *Risk Analysis: An International Journal, 32*(4), 616–632.

Liu, B. F., & Mehta, A. M. (2020, June 5). From the periphery and toward a centralized model for trust in government risk and disaster communication. *Journal of Risk Research.* https://doi.org/10.1080/13669877.2020.1773516

Mileti, D. S., & Sorensen, T. H. (1990). *Communication of emergency public warnings: A social science perspective and state-of-the-art assessment.* United States Department of Energy Office of Scientific and Technical Information Report. https://doi.org/10.2172/6137387

Nagele, D. E., & Trainor, J. E. (2012). Geographic specificity, tornadoes, and protective action. *Weather, Climate, and Society, 4*(2), 145–155.

Roberts, P. S., Velotti, L., & Wernstedt, K. (2020). How public managers make trade-offs regarding lives: Evidence from a flood planning survey experiment. *Administration & Society, 53*(4), 496–526. https://doi.org/10.1177/0095399720944811

Schwarz, M., & Thompson, M. (1990). *Divided we stand: Redefining politics, technology, and social choice.* University of Pennsylvania Press.

Slovic, P., Fischhoff, B., & Lichtenstein, S. (1980). How safe is safe enough: A psychometric study of attitudes toward technological risks and benefits. In P. Slovic (Ed.), *The perception of risk* (pp. pp. 80–103). Earthscan.

Song, G., Silva, C. L., & Jenkins-Smith, H. C. (2014). Cultural worldview and preference for childhood vaccination policy. *Policy Studies Journal, 24*(4), 528–554.

Sorensen, J. H. (2000). Hazard warning systems: Review of 20 years of progress. *Natural Hazards Review, 1*(2), 119–125.

Spennemann, D. H., & Parker, M. (2020). Hitting the "pause" button: What does COVID-19 tell us about the future of heritage sounds? *Noise Mapping, 7*(1), 265–275.

Thompson, M. (2008). *Organising and disorganising: A dynamic and non-linear theory of institutional emergence and its implications.* Triarchy Press.

Thompson, M., Ellis, R., & Wildavesky, A. (1990). *Cultural theory.* Political cultures series. Routledge.

Velotti, L. (2016). *Accountability in decision making processes for vertical evacuation adoption.* [PhD dissertation, University of Delaware]. UD Space Institutional Repository. http://udspace.udel.edu/handle/19716/21159

Velotti, L., & Justice, J. B. (2016). Operationalizing Giddens's recursive model of accountability. *Public Performance & Management Review, 40*(2), 310–335.

Wang, F., Wei, J., Huang, S. K., Lindell, M. K., Ge, Y., & Wei, H. L. (2018). Public reactions to the 2013 Chinese H7N9 Influenza outbreak: Perceptions of risk, stakeholders, and protective actions. *Journal of Risk Research, 21*(7), 809–833.

Wei, H. L., Lindell, M. K., Prater, C. S., Wei, J., & Wang, F. (2020). Texas households' expected responses to seasonal influenza. *Journal of Risk Research.* https://doi.org/10.1080/13669877.2020.1863847

Wildavsky, A. (1987, March). Choosing preferences by constructing institutions: A cultural theory of preference formation. *American Political Science Review, 81*, 3–22.

3 Research methodology

Introduction

The Italian socioeconomic scenario generated by the main consequences of the COVID-19 pandemic, coupled with the pressure that the pandemic put on the health and care system and the containment measures that were deemed necessary, was characterized by exceptional negativity, leading Italians to have a sense of fear and general concern about the aftermath of the pandemic. What worried the Italians? What has changed? And how has it impacted their daily lives? What has been, what is, and what will be their part in maintaining the processes that remain ongoing? What scenarios and what prospects are there now?

In the first place, the government actors who were managing the pandemic risk tried to answer these questions. So, too, did a host of other actors who were responsible for producing knowledge in an attempt to address and support the difficult choices that had to be made. This is how at least a hundred surveys were born within both Italian and international academies and research institutes. Among these studies, it is worth considering the national web survey on COVID-19, government measures, and the social behaviors described in this book (Lombardo & Mauceri, 2020).

The present study's survey focuses on the assessment of regulatory procedures and social behaviors undertaken by subjects in the second phase of the COVID-19 crisis in Italy. The results will help formulate risk profiles, protective behaviors, and impact assessments of dimensions such as cultural orientation and ontological (in)security that were described in the previous chapter and are operationalized here in order to create knowledge and decision-making suggestions for the future.

Research design

Previous chapters shed light on the theoretical background of and the rationale for our research, specifically the impact on people's behavior and

DOI: 10.4324/9781003187752-3

perceptions by two key factors: Ontological (in)security (i.e., one's confidence in being able to produce and sustain a shared understanding of reality) and cultural worldview. The main objective of the current research is to understand if and how these two factors may have an impact on (1) risk perception, (2) adoption of protective measures, and (3) support toward governmental regulation during the COVID-19 pandemic. Specifically, our research interest focuses on the so-called Phase 2 of the Italian national government's management of the COVID-19 crisis – a stage in which the worldview, opinions, feelings, and behaviors of Italians have been changed after being put to the test by three months of a heavy lockdown.

The research design was developed to address the research questions that emerged from the goals stated earlier:

(RQ1) Is the preference toward a specific protective behavior, or set of protective measures, affected by people's cultural worldviews?

(RQ2) What are the factors determining different degrees of ontological (in)security?

(RQ3) Did the social media infodemic influence people's trust in the ways private and public institutions have faced the COVID-19 crisis?

(RQ4) How much do people believe their private and social lives will change once the pandemic crisis ends?

(RQ5) What kind of policy and concrete actions do people believe could be implemented to manage the current situation and to put the population in the best condition to face similar crises?

The method chosen to address these research questions was an online survey, considered by our research group to be the best solution in order to collect the opinions of a wide range of respondents during the period of the pandemic. Online surveys are becoming more and more familiar to researchers as the Internet is reaching every place and every person in the world. Moreover, during the mobility restrictions forced by the lockdown measures, digital methods were almost the only feasible option to carry out social research, especially considering the research unit of analysis: Individuals aged 18 and older in Italy, a place still dealing with the COVID-19 crisis.

Research dimensions: a concept map

A combination of insights gleaned from a review of the available literature and from a research group discussion led us to select several conceptual dimensions in order to properly investigate the social behaviors undertaken by respondents during the second phase of the COVID-19 crisis.

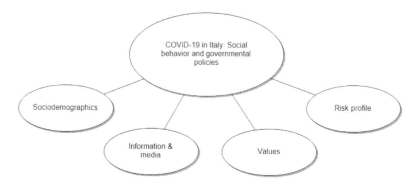

Figure 3.1 Main categories addressed in the questionnaire

The main topics addressed by our concepts are shown in the concept map in Figure 3.1.

In this map, the following concepts should be clarified:

- *Sociodemographic* refers to the characteristics of individual respondents, such as gender, age, residence, education, and so on. This section helps us to understand how much the family structure, profile, and socioeconomic status of individuals can influence the formulation of risk profiles, the emergence of certain fears, and the strength with which they present themselves, without neglecting the geographical group membership that indirectly speaks to us of the geography of the contagion.
- *Information and media* has to do with the consumption and evaluation of the sources, channels, and content of the information received during the pandemic crisis, as well as of the actors involved. This category also frames the level at which the government is perceived to have greater credibility and trustworthiness when there are inconsistencies in the received information.
- *Values* refers to cultural worldviews, political orientation, and religious beliefs of the respondents.
- *Risk profile* involves three main aspects: (1) Experience of the contagion and reproduction of the geography of the perception of the virus; (2) the protective behaviors either adopted by respondents or detected by them in their own social contexts; and (3) the degree of agreement respondents had with the regulatory provisions that were examined in various phases of the crisis, specifically to assess the measures with which the Italians were more and less likely to cooperate, while also relating this to the ontological frame built by cultural orientations and the influence they have on the degrees of acceptance of certain limitations.

The knowledge drawn from this conceptual plan will be useful in trying to understand how Italians reacted to the second phase of the COVID-19 pandemic. It will also help us in developing the rationale for managing the next phases of the emergency, to be shared with the various levels of the Italian government that are involved in crisis management.

Sampling

The reference population for this study was people living in Italy, aged 18 and over. Respondents from this reference population were involved in the research sample through a two-step sampling strategy:

1. First step: Selection of consistent and reliable online questionnaire collection points; and
2. Second step: Using the selected collection points in order to collect responses to the online questionnaire.

According to "Observatory Internet Media" (2020), Facebook is still the most widely used social media platform in Italy, and it is also the most intergenerational one. That is why the selection of the online collection points began by considering those Facebook pages and groups that met the following criteria:

- *Number and participation*. At least 500 members, each with more than 20 posts per day, used as proxy for sharing and participation in community activity.
- *Keywords*. The following keywords best represented the social media imagination with regard to the second phase of the COVID-19 pandemic crisis in Italy: Everything will be fine (#andràtuttobene), I stay at home (#iorestoacasa), at the time of covid-19 (#altempodelcovid-19), at the time of the coronavirus, emergency COVID/coronavirus, red zone, coronavirus Italy, pandemic, swabs, phase 2, masks, I do not sanction, covid2019, quarantine, revival decree, restart, lockdown, start again, new normal, post covid-19, flash mob, applaud, balconies.
- *Geographical references*. Thematic groups were selected on the basis of their link to the topics dealt with, along with the fact that they contained clear references to regions, cities, and territorial memberships (including groups of "You are from [name of the place] if . . .").

The total list surveyed includes approximately 500 pages and groups.[1] However, not all of our requests to subscribe to the selected pages and groups were approved. Neither were all posts containing survey invitations approved, reducing the number of collection points to about 300. These

Facebook groups and pages were used as collection points at which the invitation to participate in the survey was placed every week.

Considering the slowness of response and the difficulties in collecting responses from Facebook, we added other social media platforms as collection points: WhatsApp, LinkedIn, and Twitter. On top of this, we sent a request to the University of Federico II, asking them to forward to their entire organizational and student mailing list our email inviting participation in the survey.

This multiplatform procedure generated over 900 posts on pages and in groups, with at least 500 members, 100 WhatsApp contacts, six re-posts on LinkedIn and Twitter (for personal networks that count more than 1000 contacts), and more than a thousand emails sent from the university's address book.

The two-step sampling strategy led us to collect 1,267 responses. However, involving a university located in Naples caused us to have an inflated number of responses on Campania, which represented 50% of the data collected. For this reason, in our analyses, the weight of Campania is reduced by balancing it with respect to the other regions and the quotas reached for each. In other words, we have re-scaled the weights for all the regions, bringing the percentage of the regions back to the estimated value of the ISTAT (the Italian National Institute of Statistics) quota of all the residents divided by gender. This allows us to balance the effects of over- and underestimating the geographical origin, while also correcting the imbalance detected in the distribution by gender, by developing a weighting index that combines the effects of gender and origin together, maintaining the sample size at its actual value.

Considering that the link to the questionnaire may have reached nearly 50,000 users, a total response number of 1,267 may seem like a low number; that would make the response rate around 3%. However, given the willingness to participate in a survey that averaged 20 minutes to complete, the result achieved is more than adequate, since nearly all respondents completed the questionnaires through to the end. Moreover, it is noteworthy that the respondents felt involved in the research topic – a fact underlined by their responses to an optional, open-ended question at the end of the survey. Many people replied to this optional question, exceeding the usual standard of five medium-length sentences and using words that were full of emotion.

Sample description

This section illustrates the main sociographic features of our sample, and Table 3.1 sums up the distribution of respondents by gender, age, macro-area of residence, and marital status.

The sample was almost equally split between women (51.3%) and men (48.7%), with the majority of respondents aged between either 40 and 59 years (39.4%) or 29 and 39 years (30.0%). The most represented geographical areas were northwestern Italy (26.7%) and southern Italy (23.1%).

Table 3.1 Sample demographics (n=1267)

Category	Modality	*f*	%
Gender	Female	650	**51.3**
	Male	617	48.7
Age	18–28	227	17.9
	29–39	380	30.0
	40–59	499	**39.4**
	60 and over	161	12.7
Geographic area	Northwest	338	**26.7**
	Northeast	245	19.3
	Central	252	19.9
	South	293	**23.1**
	Islands	139	11.0
Marital status	Single	435	**34.3**
	Married	502	**39.6**
	Cohabitating	224	17.7
	Divorced	93	7.4
	Widow(ed)	13	1.0

Table 3.2 Distribution of frequency for education, income, occupation

Category	Modality	*f*	%
Education	Low	59	4.7
	Moderately low	418	33.0
	Moderately high	590	**46.6**
	High	199	15.7
Income	Up to €20,000	311	**24.6**
	From €20,001 to €35,000	372	**29.3**
	From €35,001 to €50,000	224	17.7
	More than €50,001	188	14.9
	Missing	171	13.5
Occupation	Unemployed	49	3.9
	Student	199	15.7
	Employed full- or part-time	658	**51.9**
	Self-employed	191	15.1
	Retired	89	7.0
	Homemaker	38	3.0
	Other	44	3.4

Table 3.1 also shows the marital status of respondents in our sample, with 36.6% married, 34.3% single, and the remaining 29.1% divided among other responses.

The general level of education of our sample was high (Table 3.2), with more than 60% of our respondents saying they have a high (15.7%) or

Table 3.3 How important is religion to your life?

Response	f	%
Not important	513	**40.5**
Of low importance	213	**16.8**
Moderately	225	17.7
Important	178	14.1
Of high importance	138	10.9
Total	1,267	100

Table 3.4 What is your political orientation?

Response	f	%
Not interested	158	12.5
Apolitical	127	10.0
Extreme right	4	0.3
Right	74	5.9
Center-right	90	7.1
Center	84	6.6
Center-left	240	**18.9**
Left	296	**23.4**
Extreme left	83	6.6
Other political movements	110	8.7
Total	1,267	100.0

moderately high (46.6%) educational level. The most represented income level per year (Table 3.2) was "from €20,001 to €35,000" (29.3%), followed by the group with an income of "up to €20,000" (24.6%). The majority of respondents were employed either full- or part-time in the public or private sector (51.9%). Religion seems not to be so relevant for most of our sample (Table 3.3): More than 57% of respondents said that religion is not very important in their lives.

As regards the political beliefs of our respondents, Table 3.4 shows there is a prevalence of left-wing positions: The groups "center-left" (18.9%), "left" (23.4%), and "extreme left" (6.6%) represent almost half of the sample.

Data collection

The survey was conducted using the CAWI (Computer Assisted Web Interviewing) method, and the responses were collected through the Google Forms platform between May 11, 2020 and June 3, 2020. The questionnaire was drawn up on the basis of the conceptual dimensions shown in the concept map (Figure 3.1), in ten sections, as listed next. For each section, the operational definition of the variables is shown in a summary table.

1. Consent agreement

The purpose of the study in which we are seeking your contribution is to understand the role played by cultural orientation in supporting the governmental restrictions put in place to contain the spread of COVID-19. In this regard, sociodemographic information and behavioral dimensions of everyday life will also be taken into account.

The project includes data collection via a CAWI (open-access online questionnaire) developed using the Google Form application.

The data collected will be used exclusively by Dr. Lucia Velotti, responsible for the study, and by Dr. Gabriella Punziano, responsible for the study's methodological aspects. The study is being conducted in both Italy and the United States of America and is completely anonymous. The total number of participants to be reached is at least 1,000 people.

Participation in the research does not entail any risk for the participant. Neither does participation in the study bring any direct benefit to the participant; although indirectly, participation in the study will contribute to the understanding of the problems related to communicating with people who have different cultural orientations.

To participate in the study, you must read and accept this information.

The questionnaire length is around 15 minutes.

If you have questions about the study, please contact the study director, Dr. Lucia Velotti, at lvelotti@jj.cuny.edu. If you have questions about your rights as a research participant, please contact the CUNY research administrator at +1-646-664-891, or by email at HRPP@cuny.edu.

---

2. Sociodemographic characteristics (gender, region, area, age, education, income, job status, job sector, household composition, and marital status; Table 3.5)
3. Risk perception, trust/confidence in governmental abilities, and adoption of protective behavior (Table 3.6)
4. Trustworthiness of sources and channels (Table 3.7)
5. Agreement with governmental measures (on work, school, commerce, social distancing, sanctions, mobility, and so on; Table 3.8).
6. Social and protective behaviors (Table 3.9)
7. Personal experiences with the COVID-19 (Table 3.10)

Table 3.5 Variables, questions, items, and measures of the questionnaire: Sociodemographic characteristics

Variable	Question	Items	Measure
Gender	What is your gender?	Male Female Other (Specify)	Nominal
Geographic area	In which Italian region do you live?	Abruzzo; Basilicata; Calabria; Campania; Emilia-Romagna; Friuli-Venezia Giulia; Lazio; Liguria; Lombardia; Marche; Molise; Piemonte; Puglia; Sardegna; Sicilia; Toscana; Trentino- Alto Adige; Umbria; Veneto; Valle d'Aosta	Nominal
Age	What is your age?	Open-ended question	Scale
Education	What is your educational level?	Some high school, no diploma High school diploma Some college credit, no degree Bachelor's degree Master's degree Doctoral degree	Ordinal
Income	What is your annual income?	Under €10,000 From €10,001 to €20,000 From €20,001 to €35,000 From €35,001 to €50,000 From €50,001 to €70,000 Over €70,000 Prefer not answer	Ordinal
Job status	What is your job?	Unemployed Student Student and full-time worker Student and occasional worker Student and part-time worker Casual worker Part-time worker Full-time worker Retired Homemaker Other	Nominal
Job sector	What is your job sector?	Public Private	Nominal
Household composition	How many people are in your family?	Newborns Children Teens Youth Adults Seniors	Scale
Marital status	Currently, you are:	Single Married Cohabitating Divorced Widow(er)	Nominal

Table 3.6 Variables, questions, items, and measures of the questionnaire: Risk perception, trust/confidence in governmental abilities, and adoption of protective behavior

Variable	Question	Items	Measure
Risk perception	In your opinion:	What is the chance that you will contract COVID-19? What is the chance that people around you will contract COVID-19? What is the possibility that COVID-19 could threaten your life? What is the chance that COVID-19 could threaten the lives of people around you?	Five-point Likert scale, from Unlikely to Likely
Perceived knowledge about Protective behaviors	In your opinion:	How well do you know how to protect yourself from COVID-19? How well do others around you know how to protect themselves from COVID-19?	Five-point Likert scale, from Not at all to Very much
Trust/ Confidence in government's ability to ensure safety	In your opinion:	How confident are you that a hypothetical new outbreak of COVID-19 will be brought under control by the national government? How confident are you that a hypothetical new outbreak of COVID-19 will be brought under control by the regional government? How confident are you that a hypothetical new outbreak of COVID-19 will be adequately managed by the Health System?	Five-point Likert scale, from Not at all to Very much
Risk perception, age-related	In your opinion:	How at risk for severe infection may infants be (0 to 1 year old)? How much at risk for severe infection may children be (2 to 10 years)? How much at risk of severe infection may adolescents be (11 to 17 years)? How much at risk of severe infection may young people be (18 to 30 years)? How much at risk of severe infection may adults be (31 to 59 years)? How much at risk of severe infection may the elderly be (60 years and older)?	Five-point Likert scale, from Very unlikely to Very likely

Table 3.7 Variables, questions, items, and measures of the questionnaire: Source and channel trustworthiness

Variable	Question	Items	Measure
Source and channel of information consumption	How much do you use each of the following channels/sources to keep informed about the pandemic?	TV news or political debate programs TV entertainment programs Radio programs of information or political debate Entertainment radio programs Scientific web or print publications Civil protection bulletin Government and institutional websites Official journalistic information websites Opinion or entertainment websites and blogs Social media (e.g., Facebook, Twitter, Instagram) Newspapers and news magazines Entertainment magazines Messaging (e.g., WhatsApp, Telegram) Request to acquaintances whom I consider experts Ask relatives/friends	Five-point Likert scale, from Not at all to Very much
Source reliability	How reliable do you consider the information you receive from:	Newspapers Newspapers on the web TV news programs TV programs of information and political debate Radio news programs Radio programs of information and political debate Posts and live broadcasts by politicians on the main social networks (fb, yt, tw, etc.) Posts and live broadcasts of government representatives (fb, yt, tw, etc.) Posts and live broadcasts of civil protection agencies (fb, yt, tw, etc.)	Five-point Likert scale, from Not reliable at all to Extremely reliable

Source consistency	Please indicate how much you agree with the following statements regarding the modalities of the reopening of the SECOND PHASE, disseminated by different levels of government:	Scientific programs and debates Posts and directives from members of the scientific community (fb, yt, tw, etc.) Informal networks (relatives, friends, acquaintances) Restrictions disseminated at the national level are inconsistent with those disseminated at the regional level. Restrictions at the regional level are inconsistent with those at the municipal level. Restrictions at the municipal level are inconsistent with those at the national level. Restrictions are inconsistent within the neighboring municipalities.	Five-point Likert scale, from Not at all to Very much
Governmental source trustworthiness/ legitimacy	When you find inconsistencies in the information you receive about the restrictions put in place by these governmental levels, whom do you trust more?	National level Regional level Local level (neighboring towns) Local level (my town) None of these	Nominal

Table 3.8 Variables, questions, items, and measures of the questionnaire: Agreement with governmental measures

Variable	Question	Items	Measure
Support for governmental policies related to school and job	How much do you agree with the following statements concerning school and work measures?	Preschools, primary, secondary, high schools, and universities nationwide must remain closed until a vaccine is found. Online learning is the most appropriate way to maintain schools' educational functions during the second phase of the COVID-19 emergency. Public and private organizations must support teleworking (smart working) during the second phase of the COVID-19 emergency. Companies must respect the ban on economic redundancies related to the COVID-19 crisis during the emergency phase.	Five-point Likert scale, from Not at all to Totally agree
Support for governmental policies related to public and commercial activities	How much do you agree with the following statements concerning public and commercial activity measures?	During the second phase of the COVID-19 crisis, e-commerce should be allowed for the same goods that retail shops can sell. Restaurants and bars should remain closed during the second phase of the crisis. Tourist services (including accommodation, ski or beach services, museums, attractions, etc.) should remain closed during the second phase of the emergency. Personal care businesses (barbers, hairdressers, beauty centers, and spas) should remain closed during the second phase of the emergency. People's temperature should be taken when entering public places, such as communities, supermarkets, and hospitals.	Five-point Likert scale, from Not at all to Totally agree
Support for governmental policies related to gatherings	How much do you agree with the following statements concerning gathering restriction measures?	The need to hold mass events (concerts, sporting events, etc.) must be carefully assessed throughout the region. All mass-gathering activities (private parties, spiritual groups, etc.) should be banned, if deemed necessary by the government, at all stages of the COVID-19 crisis. All crowded public places (not related to daily needs) must be closed, such as large shopping centers, hotels, cinemas, restaurants, and so on, if deemed necessary by the government, in all phases of the COVID-19 crisis. If deemed necessary by the government, religious services should be banned in all phases of the COVID-19 crisis.	Five-point Likert scale, from Not at all to Totally agree

Support for governmental policies related to mobility restrictions	How much do you agree with the following statements concerning mobility restriction measures?	It is necessary to control traffic and private travel of people in and out of regions. It is necessary to control national public transport traffic (trains, ships, planes, buses, etc.). Limit the entry into Italy of people coming from China's risk areas. Limit the entry into Italy of people coming from the USA. Limit the entry into Italy of people coming from Europe. Limit the entry into Italy of people coming from Africa. The southern regions should close their borders to people from the North's most affected regions (Lombardy, Piedmont, Veneto, Emilia).	Five-point Likert scale, from Not at all to Totally agree
Support for governmental policies related to self-quarantine	How much do you agree with the following statements concerning self-quarantine?	The government should call on citizens to refrain from exiting their homes unless it is an emergency. Patients with mild symptoms of COVID-19 should be isolated and treated at home, and should only contact their own doctors. All people who have returned to Italy from an area where the outbreak has occurred should undergo 14 days of home isolation and contact their doctor if symptoms are suspected. People who have been in contact with a confirmed case of COVID-19 should undergo home isolation for 14 days and contact their doctor if symptoms are suspected.	Five-point Likert scale, from Not at all to Totally agree
Support for governmental administrative sanctions	How much do you agree with the following statements concerning COVID-19 sanctions?	Violators of measures to contain the pandemic (such as wearing a mask, social distancing, only limited exit from home in case of emergencies or to reach workplaces). are subject to a fine (from €400 to €3,000). Failure to comply with quarantine measures by anyone who has tested positive for COVID-19 carries criminal penalties: imprisonment from 3 to 18 months and a fine of between €500 and €5,000, with no possibility of ablation. Violating quarantine and having contracted the virus, leaving the house, and spreading the disease may lead to prosecution for serious offenses (epidemic, murder, injury), punishable by severe penalties of up to life imprisonment. Self-certification (printed or handwritten) required in order to leave the house in case of legitimate need during the COVID-19 emergency.	Five-point Likert scale, from Not at all to Totally agree

(Continued)

Table 3.8 (Continued)

Variable	Question	Items	Measure
Support for governmental policies on financial aid	How much do you agree with the following statements concerning income protection measures?	It is necessary to consider extraordinary support measures for households with children in order to cope with the emergency caused by the outbreak of COVID-19 (such as babysitting bonuses, parental leave, etc.), at least until Phase 3. It is necessary to think about extraordinary support measures for families with elderly or disabled people (such as an increase in paid leave days under Law 104/92), to help them cope with the emergency caused by the outbreak of COVID-19, at least until Phase 3. It is necessary to think about extraordinary income support measures for workers unable to return to work (such as the redundancy fund and compensation for VAT payments), to deal with the emergency caused by the outbreak of the COVID-19 epidemic, at least until Phase 3.	Five-point Likert scale, from Not at all to Totally agree
Support for governmental policies on health protection	How much do you agree with the following statements concerning health protection measures?	Wearing masks should be compulsory for everyone. It is necessary to stay indoors to avoid creating aggregations or opportunities for contagion. It is necessary to wash hands, ventilate, and disinfect surfaces with which you come into frequent contact. I must keep a distance of one meter between myself and others when I go out, even if they are relatives living together.	Five-point Likert scale, from Not at all to Totally agree

Table 3.9 Variables, questions, items, and measures of the questionnaire: Social and protective behaviors

Variable	Question	Items	Measure
Frequency of adopting protective behavior	How often do you. . .	Wear a mask (or other respiratory protection) when you go outside? Use hand sanitizer? Wash your hands after returning home? Disinfect washable objects and surfaces? Use disposable gloves when going out? Wash clothes after returning home? Take off shoes immediately after returning home? Respect the distance of one meter between you and others when you go out? Make a self-certification when traveling among regions?	Five-point Likert scale, from Never to Regularly
Frequency of social activities	Since the PHASE TWO government restrictions, how often do you do the following activities?	Leave home for necessities (pharmacy, shopping, work, walking the dog, caring for dependent persons)? Exercise outdoors? Take public transportation? Go out for a walk? Go out with friends, relatives, or partners? Visit relatives and partners? Visit friends? Bring relatives and partners into the house? Bring friends into the house? Let outside people (domestic help, babysitters, workers) into the house?	Four-point Likert scale, from Never to Regularly
Social control measures	Since government restrictions have been in place, have you. . .	Reprimanded a stranger for not following social distancing rules? Reprimanded a family member for not following social distancing rules?	Each item is a dichotomy (Yes/No)

(*Continued*)

Table 3.9 (Continued)

Variable	Question	Items	Measure
Governmental regulations and policies impact your job	Have the measures to face the COVID-19 infection had an impact on your working conditions?	No impact I do not have a job I have lost my job I have a commercial and/or service activity and am waiting for reopenings I work from home online I am in CIG or benefit from self-employment bonuses I am on parental leave Other	Nominal
Self-efficacy	Since the measures and behaviors suggested by the government concern the whole population, do you think it would be useful for you to make a specific contribution to following these policies?	Five-point Likert scale, from Not at all to Totally agree	Ordinal

Table 3.10 Variables, questions, items, and measures of the questionnaire: Personal experiences with COVID-19

Variable	Question	Items	Measure
COVID-19 direct and indirect experience	Can you tell us if any of the following things have happened to you?	Have you been infected with COVID-19? Have you personally known someone who has been infected with COVID-19? Have you known people who know other people infected with COVID-19? Have you ever lost family members to COVID-19? Have you ever lost friends to COVID-19?	Each item is a dichotomy (Yes/No)
Social and environmental cues	How different are these things in the SECOND PHASE?	Pedestrian traffic? Car traffic? Crowding in public transportation? People wearing masks? People wearing disposable gloves? People who respect the interpersonal safety distance? People who cover themselves with their elbows when they sneeze? People who stand in public spaces (parks, benches, squares)?	Nominal variable with the following modalities: I don't know Not around me Decreased Unchanged Increased

8. Cultural worldviews (operationalized by testing a set of well-known statements; Table 3.11)
9. Political and religious orientations (Table 3.12)
10. Expectations and visions of the future (Table 3.13)

Table 3.11 Variables, questions, items, and measures of the questionnaire: Cultural worldviews

Variable	Question	Items	Measure
Cultural worldviews	How much do you agree with the following statements?	**Individualism** For the most part, succeeding in life is a matter of chance. Society would be much better off if we imposed strict and swift punishment on those who break the rules. The best way to get ahead in life is to work hard and do what you are told to do. **Egalitarianism** Our society is in trouble because we do not obey those in authority. What our society needs is a fairness revolution to make the distribution of goods more equal. Society works best if power is shared equally. Even if some people are at a disadvantage, it is best for society to let people succeed or fail on their own. **Hierarchism** We are all better off when we compete as individuals. Even the disadvantaged should have to make their way in the world. **Fatalism** Most of the important things that take place in life happen by random chance. It is our responsibility to reduce the differences in income between the rich and the poor. No matter how hard we try, the course of our lives is primarily determined by forces beyond our control.	Five-point Likert scale, from Strongly disagree to Strongly agree

Table 3.12 Variables, questions, items, and measures of the questionnaire: Political and religious orientations

Variable	Question	Items	Measure
Religious orientation	How important is religion in your life?	Five-point Likert scale, from Not important at all to Extremely important	Ordinal
political orientation	What is your political orientation?	Not interested Apolitical Extreme right Right Center-right Center Center-left Left Extreme left Other political movements	Nominal

Table 3.13 Variables, questions, items, and measures of the questionnaire: Expectations and visions of the future

Variable	Question	Items	Measure
Evaluation of institutions	How do you rate the pandemic management by institutions?	Very poor Poor Average Good Very goo	Ordinal
Future	How do you imagine your future after the end of the COVID-19 emergency?	Much improved Improved Same Worsened. Much worse	Ordinal
Emotions	The chance of being infected by COVID-19 makes you feel. . .	Depressed Angry Nervous Bored Alert Scared Irritated Anxious Indifferent Fatalist Aware and careful Not afraid Optimistic/energetic Confused Concerned about myself and others	Nominal

(Continued)

Table 3.13 (Continued)

Variable	Question	Items	Measure
Prediction of infection rate	What percentage of your community do you think will be infected by COVID-19?	Open-ended question	Scale
COVID-19: prediction of the end	When do you expect the COVID-19 crisis will be over?	May June July August September October November December At some point in 2021 or later Never	Nominal

Data analysis

Once data collection was completed, a research database was built using Google Modules; it was then converted into an IBM SPSS 23 file. Data analysis was performed in four sequential steps:

1. *Data quality.* Procedures to control the quality of the collected data were implemented to evaluate the reliability of the information contained in the dataset.
2. *Univariate analysis.* This statistical analysis explores one variable at a time; univariate analysis was carried out by our research group with a twofold objective: To describe the data, eventually finding trends that may exist within it, and to have some clues in order to transform some variables to facilitate interpretation.
3. *Bivariate analysis.* This statistical analysis explores the relationship between two variables at a time; it was performed to understand if and how results vary according to some crucial variables, such as the sociodemographic ones.
4. *Multivariate analysis.* Several multivariate statistical techniques were used to synthesize the information (i.e., the "variance" in the case of principal component analysis, or the "inertia" when dealing with multivariate techniques for nominal variables) contained in the data by simultaneously representing the variables and the cases (subjects) belonging to our dataset. Multivariate techniques are applied to more than three variables at a time and usually require at least 20 cases per

variable analyzed (Di Franco & Marradi, 2003). The multivariate statistical techniques implemented in our data analysis were selected according to the various aims and variables involved.

We applied three types of multivariate statistical techniques:

- *Principal component analysis* (PCA). PCA is applied to quantitative variables, and it is an exploratory technique at heart. "The particular nature of the technique makes it inappropriate for the measurement of relations between phenomena; on the other hand, it makes it the ideal tool for exploiting relations between lower-level phenomena in order to summarize something they have in common, that is, to measure a higher-level phenomenon" (Marradi, 1997). In other words, PCA is by far the most suitable technique when exploring a set of variables by adopting an approach based on an open dialogue with the data, that is, without rigid assumptions guiding the analysis. Starting from the analysis of the correlation matrix between the measured variables, PCA transforms a set of many quantitative variables into few representative factors or constructs. In other words, we used PCA in order to reduce a large number of variables to a smaller number of factors that were then operationalized into indexes. For each factor, the representing variables were selected by choosing those with the highest factor loadings.[2]
- *Multiple correspondence analysis* (MCA). This multivariate technique could be considered as a PCA applied to nominal and ordinal variables. As the PCA, it is a multidimensional technique, useful to extract axes of variation that would give a sense of the association among variables. One of the main differences between PCA and MCA is that the latter discriminates the variables to be put under analysis in two categories: active and illustrative. Active variables play a distinctive role in creating and defining the factors, while the illustrative variables do not, although they contribute in a relevant way to the interpretation of the factors (Delli Paoli & Addeo, 2011). We used MCA as a preliminary, but fundamental, step to perform cluster analysis procedures.
- *Cluster analysis* (CA). Finally, this multivariate approach operates more on the cases of a dataset rather than the variables. It aims at gathering the research objects into homogeneous groups in order to create typologies that could help the researchers to understand the different profiles of individuals in the sample. MCA and CA are usually used together in a sequential approach, according to the so-called French approach to data analysis developed by Benzecri (1973) and popularized in the social science community by many important works, such as *Distinction: A Social Critique of the Judgment of Taste* by French

sociologist Bourdieu (1984). Following this approach, we applied the MCA first to synthesize several sets of variables into single factors for each set, and then a cluster analysis was applied to group respondents according to the factors extracted with the MCA.

Principal component analysis was performed using IBM SPSS 23, while multiple correspondence analysis and cluster analysis were carried out using SPAD 5.5, a French software dedicated to exploratory analysis of multivariate categorical data.

Recoding of variables

This section specifies all the transformations we have applied to the variables in order to improve the data analysis and, subsequently, to facilitate their interpretation for the readers.

The variables listed in this paragraph were not originally collected with the questionnaire; they were created after the closing of the data collection, during the first stage of the data analysis structure used in this book. Generally, the transformation procedure involves a recoding of the original variables into other variables with a lesser, and then more intelligible, number of modalities.

The first set of variables concerned the *temporal dimension*. We have created the following variables:

* A dummy variable with two modalities: Before and after the reopening on May 18; and
* A variable representing the three weeks in which we have collected the data: From May 11 to 17, from May 18 to 24, and from May 25 to June 3.

It must be noted, however, that these variables are not actually shown in the following pages, because the analysis among different time groups did not provide consistent results.

A second set of variables was related to *spatial dimension*. Here is the list of the new variables in the dataset:

* The Italian regions in which respondents live were recoded according to five geographical macro-areas, defined following the ISTAT groupings: Northwest (Aosta Valley, Piedmont, Lombardy, and Liguria); Northeast (Veneto, Trentino-Alto Adige, Friuli-Venezia Giulia, and Emilia-Romagna); Central (Tuscany, Marche, Umbria, and Lazio); South (Abruzzo, Basilicata, Calabria, Campania, Molise, and Puglia); and the Islands (Sardinia and Sicily).

- A new variable, representing the risk level of the area from which the respondents hail, was created. This variable has four modalities: severity index of transmissibility at time 0 (coinciding with the rt, recorded region by region, with reference to the weekly incidence from May 4 to May 10, released on May 16); severity index of transmissibility at time 1 (coinciding with the rt, recorded region by region, with reference to the weekly incidence from May 11 to May 17, released on May 22); and severity index of transmissibility at time 2 (coinciding with the rt, recorded region by region, with reference to the weekly incidence from May 18 to May 24, released on May 29).

The third set of variables deals with the *sociodemographic dimension*. The following variables have been created:

- The age, collected as an open question, was recoded in four modalities: Young people from 18 to 28 years old; young adults from 29 to 39 years old; adults from 40 to 59 years old; and seniors aged 60 and over.
- Another variable was created by classifying the risk category of the respondents by gender and age group, following the data of the National Health Institute: Low risk for males and females from 0 to 39 years, average risk for males from 40 to 59 and for females from 40 to 69 years, and high risk for males aged 60 and over and for females aged 70 and over.
- The original answer categories of the education variable were recoded into four ranks: low=no diploma, elementary school, and junior high school; moderately low=middle school diploma and professional qualification; moderately high=bachelor's degree or professional master's degree; and high=university master's degree or PhD.
- Income was classified into a lesser number of modalities, representing four income levels: Up to €20,000, from €20,001 to €35,000, from €35,001 to €50,000, and more than €50,001).
- A variable collecting the occupational prestige in three modalities (high, medium, and low) was created by recording the occupations of the respondents.
- A socioeconomic index was created by combining educational level, income level, and employment prestige. This index has five modalities: high, moderately high, moderate, moderately low, and low.

Axial coding was adopted for open questions, which specifically concerned:

- "The possibility of getting sick from COVID-19 makes you feel _____," to which the modalities have been added: indifferent, confused, I am not afraid, fatalistic, aware, worried about me and others, or optimistic/energetic;

- The motivation for the opinion expressed on the institutional management of COVID-19 in the country; and
- The motivation for the opinion expressed on how you imagine your future after the end of the COVID-19 crisis.

Confidentiality

All respondents were guaranteed confidentiality to encourage candid responses to the research questions. This guarantee is required by the John Jay College of Criminal Justice's regulations regarding any study involving human subjects. The study provided all participants with an informed consent statement. The entire research protocol was approved by the John Jay College of Criminal Justice's Internal Review Board (IRB), with protocol number 2020–0327.

Limitations and further research

In the social sciences, all empirical studies have their methodological limitations and liabilities, whose function is not to limit or reduce the gnoseological scope of the research results but rather to frame them in the right perspective. This study is no exception.

As highlighted in the previous pages, the methodological limits of our research lie mainly in the non-probabilistic procedure with which we have built the sample, and in the narrow timeframe of our research object, Phase 2 of the pandemic in Italy.

Our sampling procedure certainly has weaknesses, such as the overrepresentation of one geographical area compared to others. According to the traditional quantitative perspective of social research, these sampling weaknesses would not allow for statistical generalization of the research results. And again, according to this line of thought, this would be a major limitation.

In recent decades, however, epistemological reflections in social research methodology have downsized the relevance of probability sampling as a necessary element to define the scientific level and the quality of research.

Furthermore, as highlighted by scholars such as Marradi (1997), the random extraction on which probability sampling is based does not absolutely guarantee statistical representativeness and, consequently, the generalizability of the results.

Shifting the focus from general considerations to what we actually did, it must first be said that the research we conducted is principally exploratory at heart, as reflected in the research questions. Moreover, the research team addressed the problem arising from the sampling procedure by adopting some statistical weighting techniques, widely used in social research, including in recent studies on the social impact of COVID-19 in Italy. (See, for example, Lombardo & Mauceri, 2020.)

With regard to the temporal scope of our research object, the opinions and perceptions of Italians who were of age during Phase 2 of the COVID-19 pandemic, there is no doubt that the results obtained from our analysis could be outdated if we consider the subsequent events that occurred in the following months.

However, any study in the social sciences must be correctly interpreted by considering the spatiotemporal context of reference, without limiting its scientific relevance and gnoseological scope (Madge, 2003). The results of our research depict a social and cultural slice of life in a Western society (Italy) grappling with an unpredictable event (COVID-19) that has created a crisis whose consequences are not yet knowable. The results illustrated in this book may be useful in interpreting other crisis situations like this one that may occur in coming years.

Moreover, our research demonstrates the validity and effectiveness of the Protective Action Decision Making (PADM) model (Lindell & Perry, 2012), in addition to the concepts of ontological insecurity and cultural worldview, in interpreting the impact on risk perception, the adoption of protective measures, and support toward governmental regulations during a crisis like Phase 2 of COVID-19 pandemic in Italy.

The research group that conducted this survey is preparing a new study in which it will try to overcome the methodological limitations set out earlier, paying attention to the sampling procedure and also expanding the nations involved, in order to adopt a comparative analysis. Furthermore, the use of the PADM will be supported by an operational definition of the variables that make up the model, and by a quantitative methodological approach that includes not only an exploratory perspective but also an explanatory one.

Notes

1 Almost 70% of these Facebook pages and Facebook groups have changed their names as the pandemic evolved.
2 Factor loadings are correlation coefficients between variables and the factors they represent. Factor loadings help in the semantic interpretation of the factors as they allow us to discriminate which variables contribute more to the formation of the factor. Variables with high factor loadings are representative of the factor, while low loadings suggest that they are not (Di Franco & Marradi, 2003; Ho, 2006).

References

Benzécri, J. P. (1973). *L'analyse de données*. Dunod.
Bourdieu, P. (1984). *Distinction: A social critique of the judgment of taste*. Harvard University Press.

Delli Paoli, A., & Addeo, F. (2011). Social network research in strategy and organization: A typology. *The IUP Journal of Knowledge Management, 9*(3), 74–97.

Di Franco, G., & Marradi, A. (2003). *Analisi fattoriale e analisi in componenti principali.* Bonanno.

Ho, R. (2006). *Handbook of univariate and multivariate data analysis and interpretation with SPSS.* CRC.

Lindell, M. K., & Perry, R. W. (2012). The protective action decision model: Theoretical modifications and additional evidence. *Risk Analysis: An International Journal, 32*(4), 616–632.

Lombardo, C., & Mauceri, S. (2020). *La società catastrofica: Vita e relazioni sociali ai tempi dell'emergenza Covid-19.* FrancoAngeli.

Madge, J. (2003). *Lo sviluppo dei metodi di ricerca empirica in sociologia.* Il Mulino.

Marradi, A. (1997). Casuale e rappresentativo: Ma cosa vuol dire? In P. Ceri (Ed.), *Politica e sondaggi* (pp. 9–52). Rosenberg & Sellier.

Observatory Internet Media (2020). *Annual report.* Politecnico di Milano.

4 The private sphere

Cultural worldviews, risk perception, and protective behaviors

Introduction

The main question that this chapter aims to answer is this: What drives/ motivates the adoption of a protective behavior? Are there different risk profiles among our respondents? To answer these questions, the micro level (i.e., the level of individual people) needs to be understood together with the social construction of COVID-19. This study conceptualizes change as being motivated by the attempt to restore situations of ontological security (Giddens, 1984), referring to one's confidence in being able to produce and sustain a shared understanding of reality.

Routine behaviors and predictability of the physical and social environment provide people with a sense of ontological security. Individuals in situations of uncertainty might experience a sense of ontological insecurity that can threaten their identity. Thus, the response to COVID-19 needs to be thought of as individuals scanning and evaluating the social and physical environment and asking questions such as, "Is there a real threat to which I need to pay attention?" This automatically leads us to the concept of risk identification, according to Lindell and Perry's model (2012), followed by the risk assessment described in Chapter 2.

The answer to this question may be very different from one person to the next. Some people might feel that they are not personally at risk but that people in their care may be; this may impact their identity as caregivers and motivate them to undertake protective action. Young people with an egalitarian worldview might comply with recommended protective actions because of their desire to contribute to their community's wellbeing. Nevertheless, the lack of compliance with protective measures, such as wearing masks and gloves, and the lack of compliance with social distancing norms might be a way to deny the existence of COVID-19 as a threat. In a nutshell, this chapter will explore the relationship between one's cultural worldviews (Song et al., 2014) and the adoption of protective behaviors.

DOI: 10.4324/9781003187752-4

The denial of COVID-19 might also be due to people's attempts to conceal their professional identity. Consider people working in the food service sector, such as restaurants or coffee shop owners. Their professional identity comprises social practices, such as shaking hands and standing in close proximity to other people – the very antithesis of the new reality determined by a pandemic. However, this does not mean that all the people working in the food service sector are unwilling to adopt protective actions. Political identity could also give rise to strands of thought that see the pandemic as a global conspiracy, fueling the relevance of no-vax or denialist organizations. Family identity could be an additional element of the ontological patchwork. Even if the data show less danger for children and young people in good health, more concern and closer adoption of protective behavior could come from mothers and fathers, while young people without caregiver responsibilities toward a child or an elderly person could adopt fewer protective measures. Moreover, the fact that COVID-19 seems less dangerous for young adults could lead them to disregard the rules and the suggested behaviors for preventing infection.

In this chapter, we explore the factors that influence individuals' adoption of protective actions by using the dimensions provided by the Protective Action Decision Making model, or PADM (Lindell & Perry, 2012). These factors are grouped into three main dimensions.

The first factor concerns *people's risk perception:* their experience with COVID-19, their reception of the warnings, and their understanding of the new daily reality shaped by the pandemic. This conceptual dimension focuses on several key points:

1. Risk perception, operationalized as people's belief that COVID-19 could be a threat for themselves and those around them, and people's belief in the likelihood of severe infection, depending upon age group;
2. Confidence or trust in themselves, in others, and in institutions, and their ability to protect themselves from COVID-19;
3. The experience with COVID-19, both direct (i.e., from having been infected) and indirect (i.e., knowing people who have been infected and having lost loved ones or acquaintances to the pandemic); and
4. Geographical proximity (measured as a proxy of the diffusion of COVID-19 to each Italian region).

The second factor deals with *people's social behaviors* adopted during COVID-19, aimed at protecting themselves and the people around them. These protective behaviors, including certain legislated behaviors, considered in this study to be new practices to be adopted, are washing hands, wearing protective masks when going out, practicing social distancing, using disposable gloves, washing clothes upon returning home, taking off shoes at home, disinfecting objects and surfaces, providing self-certification in case of movement, leaving home only with good reason, avoiding outdoor physical activity, and restricting non-family members from access to the home. This dimension also reflects on the sphere of social control, people's confidence, and their perception of external signals, understood as:

1. Reprimanding people for their lack of compliance with recommended protective behaviors;
2. Perceiving the usefulness and importance of doing one's part to improve the situation, in compliance with the rules; and
3. Paying attention to particular social and environmental cues, like noticing changes concerning social situations such as traffic, or with respect to public behaviors such as wearing a mask and disposable gloves and maintaining a safe interpersonal distance.

The last dimension focuses on *people's cultural worldviews*, operationalized by using the principles of egalitarianism, hierarchism, fatalism, and individualism (Song et al., 2014). It must also be noted that people's political, religious, family, and professional identity guide the adoption of a particular combination of protective behaviors, and here these dimensions will be used as control variables with respect to the adoption of protective behaviors by the investigated sample.

This conceptual framework has been developed in order to address the following question: With respect to the opening question on what drives/ motivates the adoption of a protective behavior, the more specific issues that will be addressed will concern what the impacts are of people's risk perception, social behaviors, cultural worldviews, political and religious orientations, and professional status on adopting protective behaviors. With respect to the opening question on the existence of different risk profiles among our respondents, the question becomes: How does this series of influences impact the concept of ontological security that theoretically guides our study?

The answers to these questions will be provided by analyzing each dimension, first separately and then looking at their interconnections in the ontological (in)security system discussed in Chapter 2, to determine how they influence people's willingness to adopt protective behaviors.

People's reception and understanding of warnings

The starting point for our analysis is individuals' private spheres and their ability to fit social, environmental, and cultural cues into a precise frame to find the best way to respond to emergencies. We aim to understand how people receive warnings and translate this knowledge into an element to be evaluated for the adoption of a particular protective behavior. In this section, we also ask ourselves what could be those elements of proximity that may justify differences among the various types of behavior people attest in our sample. These elements of proximity are three: two are related to the perception sphere, both individual and collective, and the other relates to the geographical diffusion of the virus according to the effective reproduction number (the rt index).

Regarding the first element, *individual proximity*, what we looked at in the data available to us primarily involved a difference between self-centered and other-centered risk perception, coupled with confidence in the information possessed by both themselves and others; confidence in risk management by political, administrative, and healthcare institutions; and awareness of the severity of infection for specific age groups. Respondents (see Table 4.1) seem to have been more concerned about the people around them – in terms of both the ability to contract the virus (32.9% identify low risk perception) and the threat to those people's lives (32.7% identify low risk perception) – than they were concerned for themselves (28.8% identify low risk perception) and their own lives (30.1% identify low risk perception). This was true even though the general perception was that Phase 2 was not risky, since the possibility of contracting the virus was judged to be very low. Respondents seemed confident about their knowledge regarding the options available for protecting themselves, but they were less confident about others' knowledge. This means that people felt less threatened by the virus and better able to protect themselves, while perceiving those around them as less aware of ways to protect themselves. Here we see a tendency to trust themselves and distrust others, showing a risk perception that was less self-centered and more other-centered. This comes close to what can be called "fear of the other," but in our study it shows itself more as "fear for the other" by demonstrating a greater propensity for protective behavior toward the other.

Looking again at Table 4.1, we see the dimensions of stakeholders' perceptions and people's confidence. We note that the most reliable element for our sample is health (22.8% answered "much"), concerning respondents' confidence in risk management by politicians, local administrators, and healthcare institutions. Little credit is given to the political–administrative system's ability to manage the emergency, underlining a growing level of dissatisfaction that leads to detachment from those who have managed the pandemic – an effect that remains to this day.

Regarding the awareness of people's vulnerability by age group, what Lindell & Perry (2012) call *risk assessment*, the sample shows a greater concern over the risk of severe infection as age increases, shown by considering

Table 4.1 Distribution of frequency of risk perception questions

Questions	Responses				
	Very low	*Low*	*Moderate*	*High*	*Very high*
What is the chance that you will contract COVID-19?	**28.8**	**32.7**	26.4	7.7	4.4
What is the chance that people around you will contract COVID-19?	23.0	**32.9**	31.5	9.2	3.4
What is the possibility that COVID-19 could threaten your life?	**30.1**	**34.2**	21.0	10.0	4.7
What is the chance that COVID-19 could threaten the lives of people around you?	18.3	**29.4**	26.2	18.3	7.9
How well do you know how to protect yourself from COVID-19?	1.0	7.6	25.8	**35.3**	**30.2**
How well do others around you know how to protect themselves from COVID-19?	3.8	21.3	34.1	**28.2**	12.7
How confident are you that a hypothetical new outbreak of COVID-19 will be brought under control by the national government?	13.2	**28.5**	31.8	18.2	8.3
How confident are you that a hypothetical new outbreak of COVID-19 will be brought under control by the regional government?	16.5	**24.8**	29.0	18.7	10.9
How confident are you that a hypothetical new outbreak of COVID-19 will be adequately managed by the health system?	9.7	**23.6**	33.7	**22.8**	10.2
How much at risk for severe infection may infants be (0 to 1 year old)?	**42.0**	**38.6**	9.1	6.5	3.9
How much at risk for severe infection may children be (2 to 10 years old)?	**35.9**	**46.7**	10.3	5.4	1.8
How much at risk of severe infection may adolescents be (11 to 17 years old)?	26.8	**50.1**	15.9	5.5	1.7
How much at risk of severe infection may young people be (18 to 30 years old)?	13.0	**38.8**	35.7	9.0	3.5
How much at risk of severe infection may adults be (31 to 59 years old)?	3.2	16.5	35.3	**35.3**	9.7
How much at risk of severe infection may the elderly be (60 years and older)?	1.1	1.5	19.5	26.0	**51.8**

Table 4.2 Synthetic indices of risk perception constructed through principal component analysis: Fear of COVID-19 concurrence and severity and knowledge of how to protect against COVID-19

PCA: Rotated component matrix	Fear	Knowledge
Did you contract COVID-19?	**.812**	.070
Did people around you contract COVID-19?	**.852**	.041
Can COVID-19 threaten your life?	**.777**	.030
Can COVID-19 threaten the lives of people around you?	**.831**	.041
How well do you know how to protect yourself from COVID-19?	.147	**.897**
How well do others around you know how to protect themselves from COVID-19?	−.042	**.914**

Extraction method: Principal component analysis. Rotation method: Varimax with Kaiser normalization.

the age groups of 31–59 years (35.3% answered "much") and 60 and over (26.0% answered "much" and 51.8% answered "very much"). The majority of respondents agreed that the age group 0–18 years has only a minor risk of severe infection.

To synthesize the information contained in the set of variables related to risk perception, a principal component analysis (Abdi & Williams, 2010) technique, introduced in Chapter 3, was implemented to reduce the dimensions under analysis. This analysis does not take into account the risk perception related to the severity of infection for age groups and the level of confidence in the government's ability to manage future COVID-19 waves. This is due to the low variance among the sample participants.

To an expert on disaster analysis, the synthesis that we provide next might seem simplistic; however, a reduced set of attributes is necessary to make the final risk profile more readable. The procedure involves, first, the creation of a synthetic index (by applying a PCA), and then the joining of the results by applying an MCA and a cluster analysis aimed at detecting the desired risk profiles.

Table 4.2 shows the composition of the two indices resulting from the PCA. The two indices are those related to people's fear and knowledge. To better understand their distribution and relationship with a set of structural variables, the respondents' scores on these synthetic indices were classified into three levels: High, medium, and low.[1]

Table 4.3 shows how the sample tends to position itself on average levels of concern regardless of the respondents' sociodemographic characteristics. However, looking at the distributions toward higher or lower levels of fear, we can notice that those less fearful are men, residents of central and southern Italy and the islands, not young, and with a low level of socioeconomic

Table 4.3 Levels of fear by sociodemographic characteristics

Sociodemographic characteristics		Levels of fear			
		Low (%)	Medium (%)	High (%)	Total
Gender	Female	33.1	41.7	**25.2**	100.0
	Male	**34.4**	43.3	22.3	100.0
Macro-area of residence	Northwest	29.8	39.2	**31.0**	100.0
	Northeast	28.5	42.1	**29.4**	100.0
	Central	**37.4**	42.9	19.7	100.0
	South	**33.0**	44.4	22.6	100.0
	Islands	**46.9**	46.9	6.3	100.0
Age group	Youth (18–28)	22.4	47.1	**30.5**	100.0
	Young adult (29–39)	**32.3**	45.1	22.6	100.0
	Adult (40–59)	**37.2**	39.3	23.6	100.0
	Senior (60+)	**41.8**	39.9	18.3	100.0
Socioeconomic status	Low	**56.5**	39.1	4.3	100.0
	Moderately low	**33.9**	41.9	24.2	100.0
	Moderate	**35.4**	38.9	25.8	100.0
	Moderately high	**31.2**	**45.3**	23.5	100.0
	High	25.0	**38.6**	**36.4**	100.0

status. Socioeconomic status conditions respondents' fear more than the actual ability to perceive themselves as being at risk. The same trend is confirmed when looking at the dimension of knowledge versus knowing how to protect themselves.

Table 4.4 shows that, except for those with high socioeconomic status, the rest of the sample participants felt that they had a moderately low knowledge of protective measures. It is worth recalling, at this juncture, that those who have a high socioeconomic status consist of people with a high level of education, a high level of income, and a high occupational prestige level. (See Chapter 3 for details on the construction of the socioeconomic status variable.) This means that those with a high socioeconomic status have a higher consciousness of risk because they are more educated.

Regarding *collective proximity*, the information we worked on was related to direct and indirect experience with COVID-19. Specifically, respondents were asked if they had been infected (either previously or at the time of the survey); if they knew close or distant people who were infected; and if they had lost relatives, friends, or acquaintances because of the infection. Only 5% of the sample had contracted a COVID-19 infection, giving them a direct and unmediated experience. As regards indirect experience of the disease, about 50% of the sample participants claimed to personally know people who had contracted the virus. A large majority of the sample (80%)

Table 4.4 Levels of knowledge of protective behaviors by sociodemographic characteristics

Sociodemographic characteristics		Levels of knowledge			
		Low (%)	Medium (%)	High (%)	Total
Gender	Female	**42.5**	38.6	18.9	100.0
	Male	**33.7**	43.9	22.4	100.0
Macro-area of residence	Northwest	**30.8**	45.9	23.4	100.0
	Northeast	**40.8**	39.2	20.0	100.0
	Central	**36.9**	42.9	20.2	100.0
	South	**45.1**	36.5	18.4	100.0
	Islands	**40.0**	40.0	20.0	100.0
Age group	Youth (18–28)	**45.4**	38.3	16.3	100.0
	Young Adult (29–39)	**38.4**	46.6	15.0	100.0
	Adult (40–59)	**35.7**	39.4	24.9	100.0
	Senior (60+)	**35.8**	38.3	25.9	100.0
Socioeconomic status	Low	**47.8**	34.8	17.4	100.0
	Moderately low	**39.5**	39.5	21.0	100.0
	Moderate	**38.7**	38.3	23.0	100.0
	Moderately high	**37.9**	44.9	17.2	100.0
	High	18.2	36.4	**45.5**	100.0

Table 4.5 Distribution of frequency on direct and indirect experience with COVID-19

Questions	Responses	
	Yes	No
Have you been infected with COVID-19?	5.2	**94.8**
Have you personally known someone who has been infected with COVID-19?	49.2	**50.8**
Have you known people who know other people infected with COVID-19?	**80.1**	19.9
Have you ever lost family members to COVID-19?	5.0	**95.0**
Have you ever lost friends to COVID-19?	9.4	**90.6**

said they knew people who know infected people. Only 5% of the sample had lost family members to COVID 19, and about 9.5% lost a friend.

The variables shown in Table 4.5 were then combined into an index, shown in Table 4.6, measuring the level of personal experience with COVID-19. The index of personal experience with COVID-19 was measured on a scale from 0 to 5, with 0 indicating no personal experience and

Table 4.6 Distribution of frequency of sociodemographic characteristics by levels of emotional proximity

Sociodemographic characteristics		Levels of emotional proximity			
		Low (%)	Medium (%)	High (%)	Total
Gender	Female	**52.3**	32.5	15.2	100.0
	Male	45.7	38.9	15.4	100.0
Macro-area of residence	Northwest	34.9	41.1	**24.0**	100.0
	Northeast	35.2	41.4	**23.4**	100.0
	Central	57.5	33.7	8.7	100.0
	South	**61.1**	29.4	9.6	100.0
	Islands	**67.1**	28.6	4.3	100.0
Age groups	Youth (18–28)	**61.4**	32.5	6.1	100.0
	Young Adult (29–39)	**52.2**	33.8	14.0	100.0
	Adult (40–59)	40.7	39.7	**19.6**	100.0
	Senior (60+)	50.0	31.5	**18.5**	100.0
Socioeconomic status	Low	**66.0**	21.3	12.8	100.0
	Moderately low	**53.0**	34.2	12.8	100.0
	Moderate	**52.3**	37.0	10.7	100.0
	Moderately high	43.9	36.5	**19.5**	100.0
	High	34.1	47.7	**18.2**	100.0

5 indicating the most personal experience. The scores for each respondent were then classified into three modalities: High, medium, and low. These three dimensions of personal experience were then crossed with sociodemographic characteristics. In this case, those who had a lower personal experience with COVID-19 were women, residents of the South and the islands, younger people, and those with a moderately low socioeconomic status. In contrast, residents of the North, who were older and had a medium to high socioeconomic status, showed a greater level of personal experience with the spread of the virus.

Unlike the dimension of individual proximity, *collective proximity* seems to vary more according to the respondents' sociodemographic characteristics.

The last dimension of proximity to be investigated, *geographical proximity*, was elaborated by analyzing the severity of transmissibility index (rt) during the three weeks of the survey to test whether there was a reproduction of virus perception geography consistent with the geography of the virus' spread.

Following the rt scores, an index measuring geographical proximity was built, as shown in Table 4.7. The geographical impact of the spread of the COVID infection was high in the Northwest area, moderately high in the

Table 4.7 Geographical proximity index

	Index rt incidence	Index rt incidence	Index rt incidence	Final assignment to geographical proximity index
	Week of May 4–10	**Week of May 11–17**	**Week of May 18–24**	
Northwest	High	High	High	**High**
Northeast	Moderately high	Moderately high	Moderately high	**Moderately high**
Central	Moderately low	Moderately low	Low	**Moderately low**
South	Low	Moderately low	Low	**Low**
Islands	Low	Low	Low	**Low**

Table 4.8 Fear, knowledge, and emotional proximity indices by geographical proximity indices

		Geographical proximity index				
		High	*Moderately high*	*Moderately low*	*Low*	*Total*
Fear index	*Low*	23.8	15.8	22.3	**38.1**	100.0
	Medium	24.8	18.5	20.2	**36.5**	100.0
	High	**35.1**	**23.0**	16.7	25.2	100.0
Knowledge index	*Low*	21.5	20.7	19.2	**38.6**	100.0
	Medium	29.7	18.4	20.7	**31.2**	100.0
	High	**30.3**	18.8	19.5	**31.4**	100.0
Emotional proximity index	*Low*	19.0	13.8	23.3	**43.9**	100.0
	Medium	30.8	22.4	18.8	**27.9**	100.0
	High	**41.8**	**29.4**	11.3	17.5	100.0

Northeast, moderately low in central Italy, and low in the South and the islands.

The bivariate analysis of the *geographical proximity* index with the other indices of fear, knowledge, and experience proximity shows a relationship among them: High levels of fear, knowledge, and experience proximity were linked to the geographical spread of the virus. Thus, people living in the most at-risk areas gave greater relevance to geographic proximity than those living in low-risk areas, and they showed a higher level of fear and knowledge.

Social behavior during COVID-19

This section deals with protective behaviors and the perception of social and environmental cues. The first analyzed set of variables concerns the adoption of specific behaviors and reasons to undertake certain actions.

Table 4.9 sums up the most adopted protective behaviors in the sample (in bold are the highest percentages). Results show that the following actions were regularly adopted by the vast majority (more than 65%) of respondents: Using hand sanitizer (67.7%), washing hands upon returning home (92.9%), wearing a mask (88%), respecting interpersonal distance (86.2%), taking off shoes upon returning home (71.1%), disinfecting surfaces and objects (54.3%), using gloves when going out (35.1%), and washing clothes upon returning home (23.6%). Using gloves when going out was a less-adopted behavior because of a scientific debate over the usefulness of this practice. The debate also involved a discussion of whether or not this practice could become a catalyst of infection (Yadav et al., 2020). The less-wide practice of the other two protective behaviors during the lockdown can perhaps be justified because people had not yet returned to performing all their daily activities or leaving the house regularly. (We were in a period of slow transition to normality, which we introduced in Chapter 1 as "Phase 2," and therefore respondents did not have reason to think that surfaces and objects could be infected, thus mitigating the need to disinfect them.)

Table 4.9 Frequency distribution of protective behaviors adoption

Questions	*Responses*				
How often do you currently engage in the following behaviors?	*I do not go out*	*Never*	*Sometimes*	*Regularly*	*Total*
Wear a mask (or other respiratory protection) when outside	3.2	1.6	7.1	**88.0**	100.0
Use hand sanitizer	2.2	5.6	24.5	**67.7**	100.0
Wash hands upon returning home	3.1	1.3	2.7	**92.9**	100.0
Disinfect washable objects and surfaces	2.0	7.0	**36.7**	**54.3**	100.0
Use disposable gloves when going out	3.8	**23.6**	**37.5**	**35.1**	100.0
Wash clothes after returning home	4.7	**32.3**	**39.4**	**23.6**	100.0
Take off shoes immediately after returning home	3.8	11.3	13.9	**71.1**	100.0
Respect the one-meter distance between myself and others when I go out	3.2	2.9	7.7	**86.2**	100.0
Complete self-certification when traveling between regions	18.4	17.9	17.3	46.4	100.0

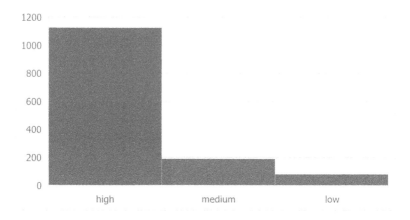

Figure 4.1 Histogram of the distribution of the adherence to protective behaviors index

The mixed results regarding the practice of self-certifying when moving between regions may be similarly understood.

A synthetic measure of the level of adherence to protective behaviors was constructed by combining the respondents' answers with the variables representing these behaviors. The scores of the index are calculated by averaging the answers: Low scores (under 1.5) are those respondents who answered with a majority of "never," so this means that medium scores (from 1.5 to 2.5) represent people with a prevalence of the "sometimes" answer; high scores (from 2.5 to 3.5) represent respondents with a high level of adherence to protective behaviors, as the majority of their answers fall into the "regularly" category.

Figure 4.1 is a histogram of the distribution of index scores. Results show an apparent propensity by the majority of the respondents to adhere to protective behaviors.

The next step was to ask respondents how often they carried out certain everyday activities before the pandemic, which are now restricted and can also be misjudged by those closely observing restraining measures (Table 4.10).

The activity most regularly carried out during the second phase of the Italian lockdown was leaving home out of necessity (about 55%). The desire to feel free from the home's narrow spaces can be seen in the responses about going outside for a walk (44% said "occasionally" + 32% said "regularly") and engaging in outdoor exercise (28% + 21%). Respondents were

Table 4.10 List of activities and frequency with respect to routines.

Question	Responses				
How often do you engage in the following activities?	*I have never done this activity.*	*I stopped this activity.*	*I occasionally carry out this activity.*	*I perform this activity regularly.*	*Total*
Leave home for necessities (pharmacy, shopping, work, dog, caring for dependent persons)	2.9	4.7	37.8	**54.7**	100.0
Exercise outdoors	32.1	19.3	**28.0**	**20.6**	100.0
Take public transportation	43.2	**42.8**	9.8	4.2	100.0
Go out for a walk	7.1	17.4	**43.7**	**31.8**	100.0
Go out with friends, relatives, partners	16.2	**35.8**	33.8	14.1	100.0
Visit relatives and partners	13.6	**28.5**	**39.8**	18.1	100.0
Visit friends	22.9	**48.5**	21.1	7.5	100.0
Bring relatives and partners into the house	16.5	**31.5**	**34.4**	17.7	100.0
Bring friends into the house	25.9	**48.7**	16.7	8.6	100.0
Let in external people (domestic help, babysitters, caregivers)	34.8	**42.3**	14.4	8.6	100.0

unwilling to abandon activities such as visiting parents and other relatives (28.5% of people declared a cessation of this activity, compared with 39.8% who said they carried out this activity occasionally), as well as bringing relatives and parents into their own home (31.3% have stopped doing so and 34.5% do it occasionally). The activities that were completely stopped correspond to the most accepted restrictions. These activities are: taking public transportation (43%); going out with friends, parents, and other relatives (35.8%, justified by a not irrelevant weight of the youngest component in the sample); visiting friends (48.5%); bringing friends into their homes (49%); and admitting outsiders (42%).

This set of variables was also combined in an index representing the level of adherence to restrictive measures by using the same procedure implemented for the adherence to protective behaviors index, assigning low scores of adherence when the respondent answered that he or she does not go out or never adopts that behavior, a medium level of adherence when the option "sometimes" was chosen, and a high level of adherence when the answer most chosen was "regularly." Figure 4.2 shows the distribution of

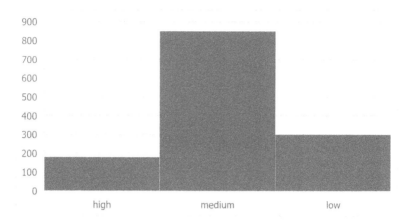

Figure 4.2 Histogram of the distribution of the index of levels of adherence to restrictive measures

the index with a propensity of the sample to be afraid to follow the restrictive measure if not also adopting a very strong adherence profile.

The data we have just reviewed give us an image of a sample that is particularly attentive to the adoption of protective behaviors, especially normative (such as wearing masks or keeping a safe distance, rather than not going out with or welcoming friends into their homes). To a lesser extent, restrictions were accepted when visiting relatives and family members, or when taking a walk or engaging in outdoor exercise. This particular datum likely denotes the great importance of the parental–familial dimension that the lockdown itself failed to undermine, together with an understanding of the importance of assuming protective behaviors accepted in a more or less generalized manner.

When it came to analyzing the environmental and social context, we worked on two kinds of features that Lindell and Perry (2012) identify: Environmental and social cues. "Environmental cues are sights, smells, or sounds that signal the onset of a threat whereas social cues arise from observations of others' behavior" (Lindell & Perry, 2012, p. 2). Thus, we asked our respondents if they noticed changes concerning foot and car traffic, crowded public transportation, or variation in the behavior of others, like wearing a mask and disposable gloves, maintaining a safe interpersonal distance, and stopping in public spaces.

Recalling that the survey period has covered the second phase of lockdown, with its gradual increase in reopening and the loosening of

Table 4.11 Awareness of change in environmental and social cues from Phase 1 to Phase 2 of COVID-19 crisis management

Question	Responses					
Compared to this list of situations and compared to Phase 1, what kind of variation are you noticing in this second phase?	I do not know.	It is not a behavior adopted by people around me.	Reduced	Unchanged	Increased	Total
Pedestrian traffic	3.4	0.8	21.0	7.4	**67.4**	100.0
Car traffic	3.0	0.5	22.9	7.3	**66.3**	100.0
Crowding in public transportation	**34.5**	9.2	27.3	8.6	20.3	100.0
People wearing masks	2.6	0.4	20.8	32.6	**43.6**	100.0
People wearing disposable gloves	7.9	1.3	**30.2**	28.7	31.9	100.0
People respecting the interpersonal safety distance	3.5	0.7	**36.4**	**32.8**	26.6	100.0
People sneeze while covering their face with elbows	38.1	7.2	10.2	**26.8**	17.7	100.0
People standing in public spaces (parks, benches, squares)	15.4	3.0	15.7	12.7	**53.2**	100.0

restrictions, what immediately catches the attention is that the results reveal an increase in the perception of pedestrian and car traffic, with a related perceived increase in people who stop in public spaces, which is an environmental cue that suggests a decrease in general concern for the spread of the virus in this specific phase. However, the loosening of containment measures does not correspond to a decrease also in the adoption of protective behaviors, such as wearing a mask (44% of the sample said they noted an increase in noticing people around them adopting this behavior). At the same time, sneezing into the elbow to avoid the spread of possible infectious particles remains unchanged. A decrease was reported, however, in the adoption of behaviors like maintaining social distancing (which had a decreased perception of 36.4%) and using gloves (which decreased in the perception of study participants by 30.2%).

The situation of crowding on public transportation is not known, as many people continue to choose private transport options and are not aware of what is happening on public means of transportation.

In an environment characterized by high uncertainty, the sample shows a heightened sense of alert to environmental and social signals as a way to make sense of the surrounding environment. Thus, one might ask how much this attention to signals is then converted into forms of social control, as

Table 4.12 Frequency distribution of social control toward relatives and strangers

Question	*Responses*		
Since government restrictions have been in place, have you:	*Yes*	*No*	*Total*
Reprimanded a stranger for not following social distancing rules?	31.3	**68.7**	100.0
Reprimanded a family member for not following social distancing rules?	**50.7**	49.3	1000

Table 4.13 Distribution of frequency of the perception of self-efficacy

Question	*Responses*	*%*
Given that the measures and behaviors suggested by the government affect the entire population, do you think it is helpful for you to follow these directions?	Do not agree at all	3.7
	Somewhat agree	1.7
	Agree	13.7
	Agree very much	**18.5**
	Completely agree	**62.4**
	Total	100.0

well as perception of the usefulness of individual contribution to the management of the pandemic phenomenon.

The dimension of social control was analyzed by asking if respondents reprimanded relatives, friends, or other people who were not engaging in one of the most critical protective behaviors. Data showed that more than 50% of the sample had blamed a family member for not following social distancing rules. This result could be explained by both familiarity and a greater level of concern for family members, which was not the case when it was a stranger who was breaking this rule (about 69% did not reprimand).

As regards their perception of the usefulness of doing their part to improve the situation by complying with the rules for containing the spread of COVID-19, most respondents agreed that they perceived themselves as doing their part (62.4%).

Cultural worldviews, political orientation, and religious influence as values to cope with COVID-19

If, up to this point, we have analyzed the influence of the individual sphere on respondents' behaviors by considering the levels of perception, proximity, and achievement from environmental and social cues, in this section we will address the sphere of influence in the adoption of protective behaviors more connected to the individuals' *values*. In particular, cultural worldviews (Song et al., 2014), political orientation, and religious influences were evaluated to understand how people cope with the threat of COVID-19.

The cultural worldviews were defined following the results of a study from Song et al. (2014) in which the authors affirm that personal values and beliefs exert a critical influence on the understanding and reaction in a risk situation. Worldviews are understood as personal values and beliefs that derive from the co-construction between the individual, the society in which he/she lives, and the set of traditions and orientations that constitute the most intrinsic substratum of the cultural worldview. Worldviews could then be considered as those principles that guide the processes of action, interaction, and reaction that, in daily life, individuals are called upon to put into practice and, even more so in moments of systemic crisis, govern the adoption of certain procedures instead of others. It is from this consideration that we have included in our web survey a set of questions intended to collect information about the four cultural worldviews highlighted in the study mentioned earlier: Egalitarianism, hierarchism, fatalism, and individualism. These four windows into cultural orientations provide insights into the orientations of our sample.

Table 4.14 shows the result of a principal component analysis applied to interpret the information shared by the variables under analysis to extract

Table 4.14 Principal component analysis of the cultural worldview set of variables

Rotated component matrix

Items	1 Egalitarianism index	2 Hierarchism index	3 Fatalism index	4 Individualism index
For the most part, succeeding in life is a matter of chance.	−.006	.086	**.775**	−.003
Society would be much better off if we imposed strict and swift punishment on those who break the rules.	−.065	**.792**	.137	.048
The best way to get ahead in life is to work hard and do what you are told to do.	.078	**.811**	−.040	.127
Our society is in trouble because we do not obey those in authority.	.082	**.764**	.176	.092
What our society needs is a fairness revolution to make the distribution of goods more equal.	**.827**	.078	.095	−.099
Society works best if power is shared equally.	**.732**	.028	.082	−.001
Even if some people are at a disadvantage, it is best for society to let people succeed or fail on their own.	−.072	.056	.017	**.847**
We are all better off when we compete as individuals.	−.076	.187	.134	**.783**
Even the disadvantaged should have to make their way in the world.	.660	.064	−.064	.000
Most of the important things that take place in life happen by random chance.	.121	.026	**.785**	.216
It is our responsibility to reduce the differences in income between the rich and the poor.	**.754**	−.076	.044	−.082
No matter how hard we try, the course of our lives is primarily determined by forces beyond our control.	.025	.135	**.746**	−.014

Extraction method: Principal component analysis. Rotation method: Varimax with Kaiser normalization.

those components that synthesize the most significant part of the total vari-ance and the greatest number of variables with high loadings.

In line with the research of Song et al. (2014), all variables are represented and have high factor loadings with the component to which they should belong. In other words, the principal component analysis has extracted four components that represent the four cultural worldviews, underlining the validity of this model for interpreting another cultural context, Italy, which is very different from the Amerindian one on which the authors' interpreta-tion is based. The *egalitarianism* dimension is represented by the variables "what our society needs is a fairness revolution to make the distribution of goods more equal"; "society works best if power is shared equally"; and "it is our responsibility to reduce the differences in income between the rich and the poor." For the *hierarchism* dimension, we found variables like "society would be much better off if we imposed strict and swift punish-ment on those who break the rules"; "the best way to get ahead in life is to work hard and do what you are told to do"; and "our society is in trouble because we do not obey those in authority." For the *fatalism* dimension we have "for the most part, succeeding in life is a matter of chance"; "most of the important things that take place in life happen by random chance"; and "no matter how hard we try, the course of our lives is largely determined by forces beyond our control." And finally, for the *individualism* dimension, we have the variables "even if some people are at a disadvantage, it is best for society to let people succeed or fail on their own"; and "we are all better off when we compete as individuals."

The only variable whose effect was not discriminating for any of the four constructed components is "even the disadvantaged should have to make their way in the world." This may be because this variable is most subject to the social desirability of the response to which it comes.

Looking at the indices' distributions (Figure 4.3), we can say that the sample is more oriented toward egalitarianism; it has a normal bell-shaped distribution concerning the propensity for hierarchism and shows a lower adherence to fatalism and individualism. So that we could reuse these indi-ces in subsequent analyses, they were also normalized and recoded into variables with three modalities: Low, medium, and high.

Political orientation and the influence of religion in one's life were then analyzed. As regards the political dimension, the sample is distributed as follows: 35% identified as "disinterested/apolitical," 13.3% as "right-wing (from the extremes to the center)," 6.6% as "center," 48.9% as "left-wing (from the extremes to the center)," and 8.7% as "belonging to other move-ments and formations."

The analysis of the importance in one's life as attached to religious faith shows how Italy remains a country in which Catholicism is deeply rooted,

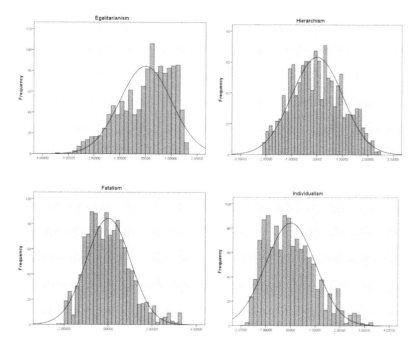

Figure 4.3 Histograms of the distribution of the four cultural worldviews components among the sample participants

to the point of finding that 11% responded "very important," 14% "quite
important," 17.7% "moderately important," 16.8% "not very important,"
and 40.5% "not important at all." Rather than evaluating the four principal
components described in this section, Table 4.14 will be used in an exploratory multidimensional analysis. This technique is aimed at producing perceptual and behavioral profiles with all the different interpretations that
have been covered, while also recovering the dimension of professional and
family composition discussed in the previous chapter.

Ontological effects on the undertaking of protective behaviors

The procedure described next was developed for exploratory purposes.
It uses a sequential combination of multidimensional analysis techniques
that are aimed at reducing the amount of information into fewer dimensions, latent variables, components, or factors. These are extracted through

multiple correspondence analysis and subsequently used for determining homogeneous classes of individuals. This is accomplished through the cluster analysis procedure that was explained in Chapter 3, which aims at producing perceptual and behavioral profiles.

In order to make this procedure operational, a matrix was built containing the main sociodemographic characteristics of the sample (sex, age group, macro-area of residence, socioeconomic status, occupational status, and family type), together with the indices of fear, knowledge, experience proximity, geographical proximity, a reprimand to strangers or family members, perception of the usefulness of one's behavior, the importance of religious belief, political orientation, cultural worldview, level of adherence to the adoption of protective behavior, and level of adherence to respect for the restrictive measures. As mentioned previously, the indices were operationalized in ordinal variables with three levels: High, medium, and low.

This statistical approach was developed in order to answer the following research questions:

(RQ1): What factors from the sphere of private influences were involved in determining the frames that guide the adoption of protective behavior in the sample investigated?

(RQ2): To which specific ontological forms did they lead us?

The first question was addressed through the application of multiple correspondence analysis procedures.

All variables relating to systems of influence in the individual sphere, reproduced here as indices, were used as active variables (i.e., directly determining the factors). All of them, except for political orientation (because of the skewedness of its distribution toward the left), caused a significant number of problems of synthesis in the procedure. Therefore, this and all sociodemographic variables were employed as explanatory variables, useful in better defining the synthetic factors extracted and, later, in describing the groups of ontological forms we looked at.

The two factors resulting from the application of this procedure, and leading us to extract about 55% of the total inertia contained in our database, can be described as follows:

• The first factor (horizontal axis, 30.2% of inertia explained): *Responsibilities vs. Liabilities.* On this factor, a system of opposition emerges between a proactive behavioral dimension, in which the belief and value system plays a key role in inducing the adoption of protective behavior beyond the fact that it becomes the subject of regulatory containment measures (left-hand side), and a passive behavioral dimension in which beliefs and values lose their significance, and lack of perception of the importance of doing one's part in combating the pandemic leads to formal adherence to restrictive measures without, however, adherence to the responsible adoption of protective behavior (right-hand side of the plan).

• The second factor (vertical axis, 24.3% of inertia explained): *Rational Pragmatist Approach vs. Deterministic Fate-Led Approach*. This factor represents the opposition between the various approaches to risk perception. In the upper part of the plan, a rational and pragmatic component prevails, based on medium-high levels of knowledge, awareness, and perception of geographical and experiential proximity, together with a collectivist and egalitarian cultural approach. On the other hand, in the lower part of the plan, there is little knowledge of how to protect oneself, little knowledge of the danger of proximity, and a value system tending toward fatalism that considers the events of the world as governed by a predetermined destiny, already established and therefore subject to social determinism.

The two factors thus constructed were cross-referenced with each other to create a factorial space of attributes in which to project the position of all the modalities of the active and illustrative variables used in the analysis. Figure 4.4 shows this plan. However, before commenting on the connections produced in this plan, in order to further facilitate the synthesis of the characteristics investigated, a further procedure was developed to classify the respondents into homogeneous groups that would allow us to speak of the distribution of characteristics in the plan, presented as combinations that lead to the definition of specific groups. And this leads us to directly address the second research question. Through the application of semi-hierarchical cluster analysis, three groups were identified that summarize respectively 38%, 26%, and 36% of the variance of the individuals in the sample.

In the upper-left quadrant of Figure 4.4, obtained from the intersection of active, responsibilities-led behavior and a rational, pragmatic approach to risk perception, we can identify the first group, labeled medium ontological insecurity (accounting for 38% of the sample). This is a group in which characteristics such as the medium-high perception of the virus' experiential proximity, medium knowledge and awareness, and a high perception of fear and self-efficacy are associated, so much so that this group presents a strong adherence to the adoption of protective behavior, to the detriment of low adherence to restrictive measures, merely because other social actors regulate these. The space of medium ontological insecurity is governed by a low individualism and a high egalitarian conception, all elements that are found in the implementation of proactive behavior and is also expressed in the action of reprimanding strangers or family members if they are caught in the act of contravening the rules that currently dominate the scenario of securing against the possibility of contracting COVID-19. This group is mainly made up of elderly (over 60 years old) workers or retired people, who live as couples, are married, or take care of single-parent families as widow(er)s; they have a socioeconomic status tending toward high and a political orientation that moves through the left. This is a group of people who are highly responsible for themselves and others, both near and far. For these people, a low ontological insecurity is given by the 60-years-old-and-over category, considered by most sources of information to be the prominent risk category. The people belonging to this group perceive, accept, understand, and process risk. In responding to COVID-19, people with low ontological insecurity chose to adopt moral behaviors that were not intended to be regulated or sanctioned, but that are part of the responsibility of each individual to society as a whole.

Moving on to the two quadrants on the right-hand side of Figure 4.4, a larger group of characteristics unfolds across the half-plan characterized by an orientation to passive behavior, in which a rational and more fatalistic component prevails in a mixed way, leading us to define this group as a space of medium high ontological insecurity (26% of the sample). In this group, characteristics such as moderately high geographical proximity of the virus, high self-efficacy, and the perception of knowing how to protect oneself or of knowing the ability of others to protect themselves from the virus, coexist with a decline in the importance of values such as religion and the influence of the authoritarian component. This latter behavior is typical of this group, which declares a low adherence to the adoption of protective behaviors and a high level of adherence to restrictive measures because they are regulated, thus underlining the characteristic liability component of this group. They still showed little concern, despite the high geographical proximity and intimate knowledge of the phenomenon, which seems to delimit a mechanism that leads to the acceptance of a situation of risk, almost ignoring it and remaining passive. And this is even more evident for the adherence to a high level of fatalism that guides their lives. They are primarily men, residing in northern or central Italy. They tend to be adults, single or cohabiting, separated or divorced, and self-employed. And they identify with other movements or political formations and have a high or moderate socioeconomic status. Compared to the previous group, members of this group do not have families with children, so the caregiver component is missing; they also have a high propensity for individualism. This group has a high risk tolerance. The propensity to accept a high level of risk might be driven by professional motivations that counterbalance risk perception.

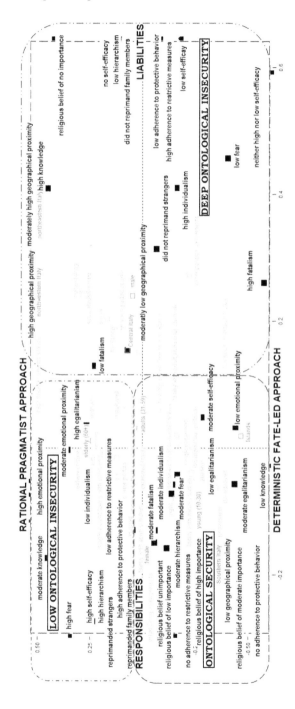

Figure 4.4 Factorial plan generated through multiple correspondence analysis on COVID-19's influence factors, with the projection illustrative of sociodemographic characteristics of the sample. Inertia explained using Benzécri correction factor: First factor 30.2%, second factor 24.3%. Semi-hierarchical clusters with 3-class cutoff respectively of 38%, 26%, and 36% variance reproduced. Elaboration with SPAD-COHERIS.

The last group can be found at the lower-left, in the intersection of a deterministic, fate-driven approach to risk perception, with a prevalence of proactive asset-led behavior. The group that falls into this space has been termed ontological security (36% of the sample). Poor or low knowledge, awareness, experiential perception, and geographic proximity of the virus diffusion prevail in this group. Also, people in this component scored average on values such as hierarchism, egalitarianism, individualism, and fatalism. Religion's influence on the people in this group ranges from low to high levels, and the perception of the importance of their behavior is defined as "enough." This group is less characterized by expressing itself with non-adherence to protective measures and non-adherence to the adoption of protective behaviors to safeguard themselves from the pandemic threat, by which they do not seem to feel threatened. This group includes the youngest members of the sample and is made up of adults, singles, and families with children; they are people with a lower socioeconomic status, sometimes students or unemployed people or housewives, who go from showing a tendency toward disinterest in politics to a timid positioning to the right. They reside, for the most part, in the South or in the islands, which, if on the one hand, justifies a more distant perception of the risk (given that, in the initial phase of diffusion of COVID-19, Italy saw the northern areas of the country most affected), does not justify the low level of concern. This is all the more so given the fact that many of the respondents are also parents; for them, therefore, the caregiver component can be decisive. This shows, in a way that leaves little room for doubt, how moderately high situations of ontological insecurity are primarily driven by the professional component. Meanwhile, ontological security, where one does not perceive the actual risk, is affected by family composition and socioeconomic status. It is therefore also connected to relevant educational poverty and open scenarios of contextual possibility, largely because this group is concentrated in the South, which has notoriously been affected by various gaps and systems of inequality.

All of these reflections on influences on the private sphere and the formation of interpretive and reactive ontological frames in our sample open our thinking in the direction of a new question: What happens when the effect of the public sphere and other influences overlap with the private sphere? We will attempt to answer this research question in the next chapter.

Note

1 The procedure involved, first, a normalization of the component scores ranging from 0 to 100 by subtracting from the individual index score the minimum value of the distribution, then dividing it by the range of the variation of the scores and multiplying the result by one hundred. Subsequently, the normalized score was divided into the three categories mentioned earlier.

References

Abdi, H., & Williams, L. J. (2010). Principal component analysis. *Wiley Interdisciplinary Reviews: Computational Statistics, 2*(4), 433–459.

Giddens, A. (1984). *The constitution of society: Outline of the theory of structuration.* University of California Press.

Lindell, M. K., & Perry, R. W. (2012). The protective action decision model: Theoretical modifications and additional evidence. *Risk Analysis: An International Journal, 32*(4), 616–632.

Song, G., Silva, C. L., & Jenkins-Smith, H. C. (2014). Cultural worldview and preference for childhood vaccination policy. *Policy Studies Journal, 42*(4), 528–554.

Yadav, D. K., Shah, P. K., Shah, S. P., & Yadav, A. K. (2020). The use of disposable gloves by general public during COVID-19 increases the risk of cross-contamination. *Asia Pacific Journal of Public Health, 32*(5), 289–291.

5   The public sphere

Risk communication, support
for public policies, and
trustworthiness of sources and
channels

Introduction

The public sphere is the key dimension of analysis in this chapter. We have conceived it as the set of exogenous factors that might influence how individuals have rebuilt and reshaped their worldview during the COVID-19 pandemic. Unlike the individual sphere, which is largely based on the concreteness of relationships with others and with the surrounding environment, the public sphere is a more abstract system, in which the processes of information and risk communication and the actors responsible for managing these processes both play fundamental roles. The public sphere dimension will be analyzed under the theoretical framework developed by Lindell and Perry's (2012) PADAM model, which has once again proved useful in interpreting how individuals' decisions toward a risky situation (such as a pandemic) may be influenced by elements linked to the institutional communication and media system.

Information sources and both channel access and preference are two elements brought back into Lindell and Perry's scheme as reinforcement and amplification of those warning signals that are expressed through the message of the wider public sphere. This configuration implicitly recalls, on the one hand, the concepts of use, trustworthiness, and consistency of information and risk communication processes and, on the other hand, compliance with pandemic public policies – in other words, decisions of whether or not to give support to the measures of contrast, containment, and social protection put in place by institutions after the first effects of the pandemic on the Italian socioeconomic and health systems.

The analysis of the survey results focuses on COVID-19-related news consumption made by our respondents. It addresses the following research questions: What channels are considered the most reliable and trustworthy when getting informed about the pandemic? Which level of government do people consider the most reliable when the information released is conflicting?

DOI: 10.4324/9781003187752-5

Many and different communication channels implemented risk communication during the pandemic; this situation has created a cacophony of increasingly unstable and undistinguishable sources that may have generated different levels of comprehension of the main topic related to the COVID-19 emergency in our sample and may have impacted the trust that people had in the credibility of the institutional actors.

Before the Internet revolution, risk communication was a prerogative of the mainstream media; now it has become the direct content of Internet-based channels and social platforms. An example of this is the use of Facebook and YouTube for daily and weekly updates on the progress of the pandemic, institutionally broadcast by the Italian Istituto Superiore di Sanità (ISS – National Institute of Health System) as a television program. The same multi-channel broadcasting approach was applied to the live television/social media communications of Prime Minister Giuseppe Conte, and also to the all-consuming information and communications that were disseminated from wherever possible to reach the largest and broadest population – generating, in the process, uncontrolled pervasive effects, like anxieties, fears, and the urge to have all the information centered on the COVID narrative. Thus, this narrative became not only the prevailing content of public discourse but was also discussed, shared, and re-elaborated by individuals who live in an information bubble entirely focused on COVID.

Everybody seemed to have something to say when the topic of discussion was COVID-19. Since different and disparate perspectives existed on infectious disease containment, prevention, and recovery measures, we asked our respondents about their agreement with the measures put in place by the government. And we cannot overlook the influence that this new, all-pervasive communication sphere can have on social media, which acted as "plague spreaders," disseminating fake news, hoaxes, and conspiracy theories and spreading uncertainty, anguish, and anger among Italian citizens. The social media infodemic (Cinelli et al., 2020) has exacerbated the health emergency (Mesquita et al., 2020); according to some authors, management of the crisis in the first phase of the pandemic resulted in many errors, some caused by the proliferation of fake news (Ruiu, 2020). The increase of fake news generated a widespread distrust toward information conveyed by institutions and the media, both traditional and digital; the only exceptions were communications from national security bodies and representatives of the scientific community considered reliable by our respondents. Moreover, the condition of isolation caused by the spread of COVID-19, and the direct effect of the national lockdown, have changed people's daily habits regarding time, space, activities, social relations, and more.

The Internet had a clear and decisive impact on the reconfiguration of these processes. In particular, online communication played a central role in creating and consolidating dynamics and practices of sociality experienced mainly in the virtual environment, given the impossibility of realizing them in the typical physical contexts of everyday life. However, these important issues related to misinformation, infodemics, and the consideration of the Internet as a transposed scenario of sociality for newfound closeness and sharing, despite the limitations of the physical presence imposed by the spread of the virus, do not find space in these pages because they are difficult to investigate through a very detailed survey, such as the one we proposed to our sample. The hope is to continue reflecting on these issues and to find spaces for analyzing the content disseminated on the web during the period of investigation considered, and specifically relating to the second phase of the management of the pandemic emergency in Italy (De Falco et al., 2021a, 2021b).

In this chapter, we will discuss the data collected on communication channels, information sources, the reliability and credibility of the institutional and non-institutional actors, the dimensions of co-governance, and the possible perception of incongruity, regulatory measures, and the forms of acceptance to which they are subject. All of this will lead to the definition of different indices that can enrich the scheme of ontological dimensions introduced and discussed in Chapter 4, resulting from the different impacts that the public and private spheres can have on the adoption of certain protective behaviors and the formation of specific risk profiles in our investigated population.

Message, sources, and channels of information use, trustworthiness, and consistency in institutional responses to COVID-19

In this study, we conceive the public sphere as the set of external influences that affect individuals' ability to form an overall picture of what is happening in the risk situation generated by the COVID-19 crisis. In contrast to the social and environmental cues discussed in the previous chapter, which in some ways are based on the assumption of proximity between events/ elements and perceived conditioning, the public sphere refers to a more abstract system of interconnections whose pivotal point is the processes of risk communication and the actors who are responsible for managing these processes. This section will discuss the systems of trust and credibility given to the sources and channels through which the pandemic information is disseminated, and the trust conferred to the actors in the three different levels of governance that are involved.

The first dimension investigated was the use, during the second phase of COVID-19 management, of some information channels by respondents to keep them updated on the pandemic. The results shown in Table 5.1 highlight that, in uncertain times and with every type of public and media debate focused on COVID-19, mainstream media such as television, radio, and traditional print press become less-used sources when compared with bulletins of civil protection, official information from the bodies in charge of pandemic management, scientific publications, and official journalism on the web, showing that the difference is no longer in the channel used but in the reliability of the source. This also explains why a minority of respondents considered social media, peer-to-peer messaging systems, and the conferral and direct requests to acquaintances, friends, and relatives, while what

Table 5.1 How much do you use each of the following channels/sources to stay informed about the pandemic?

	Very little	*Little*	*Medium*	*Much*	*Very much*
TV news or political debate programs	**25.2**	**24.5**	23.3	13.3	13.7
TV entertainment programs	**69.2**	**15.8**	9.2	4.2	1.6
Radio programs of information or political debate	**48.5**	**26.9**	14.0	6.3	4.3
Entertainment radio programs	**69.6**	**19.7**	6.0	3.9	0.8
Scientific web or print publications	10.4	19.4	26.7	**23.5**	**20.1**
Civil protection bulletin	12.8	19.2	26.2	**20.4**	**21.5**
Governmental and institutional websites	11.2	21.1	22.5	**23.6**	**21.6**
Official journalistic information websites	9.4	15.9	26.3	**28.7**	**19.7**
Opinion or entertainment websites and blogs	**47.0**	**27.1**	14.6	8.1	3.2
Social media (e.g., Facebook, Twitter, Instagram)	**30.9**	**24.1**	22.8	12.9	9.3
Newspapers and news magazines	**37.0**	**24.5**	19.2	12.5	6.8
Entertainment magazines	**80.8**	**13.2**	4.0	1.6	0.4
Messaging (e.g., WhatsApp, Telegram)	**52.9**	**20.5**	12.6	7.5	6.6
Request to acquaintances whom I consider experts	**45.4**	**21.0**	17.5	11.1	5.1
Ask relatives/friends	**58.5**	**24.1**	10.7	5.1	1.7

were more sought after were informative sources that could be transmitted in a multi-channel manner (e.g., live on television and social media, or as content on official governmental and institutional portals that can be used on different digital platforms).

While the decreasing consumption of mainstream channels is already established in the literature with regard to the information channels used during disasters and pandemics (Reynolds & Seeger, 2005), the results of the increased narrowness of network and social media use seem quite discordant with recent studies (Wendling et al., 2013). Therefore, it becomes useful, having noted the difference in source and channel use, to understand what confers greater reliability and trust (information shown in Table 5.2). Once again, there is little ability to trust the classically understood media, with its one-to-many communication model, regardless of the source of the message. Even social media and digital platforms are not trusted if politicians or government representatives communicate through them, while levels of trust increase if entities such as the civil protection agency and the

Table 5.2 How reliable do you consider the information you receive from these channels to be?

	Very little	Little	Medium	Much	Very much
Newspapers	9.0	**24.5**	**51.1**	13.6	1.8
Newspapers on the web	8.4	**32.1**	**48.6**	9.2	1.7
TV news programs	9.7	**21.4**	**46.6**	17.9	4.4
TV programs of information and political debate	16.2	**29.6**	**41.2**	12.1	1.0
Radio news programs	9.5	**25.4**	**48.5**	15.2	1.4
Radio programs of information and political debate	12.9	**30.4**	**42.3**	13.4	1.1
Posts and live broadcasts by politicians on the main social networks (fb, yt, tw, etc.)	**35.1**	**39.3**	18.6	4.8	2.1
Posts and live broadcasts of government representatives (fb, yt, tw, etc.)	**21.3**	**29.0**	**33.1**	10.9	5.7
Posts and live broadcasts of civil protection agencies (fb, yt, tw, etc.)	10.8	**20.7**	**33.4**	**23.5**	**11.6**
Scientific programs and debates	4.3	12.7	**33.8**	**33.1**	**16.1**
Posts and directives from members of the scientific community (fb, yt, tw, etc.)	7.2	14.2	**31.3**	**32.5**	**14.8**
Informal networks (relatives, friends, acquaintances)	**44.3**	**35.6**	16.0	3.1	1.0

community of scientists (who are also accredited through traditional channels and not only on the Internet) communicate on digital social platforms. At the lowest levels of reliability and trust are positioned the informal networks constituted by acquaintances, friends, and relatives, about which there is probably fear of the accumulation of inaccurate news in the midst of the general information chaos that characterized the spread of news about COVID-19.

The results highlight three oppositions among sources: The first contrast is between scientific sources and other sources; the second opposition occurs between politicians/government and civil protection agencies; and the third opposition is between unreliable information and scientific information.

In order to deepen the dimension of trust in institutions and governmental agencies, two more questions were posed to the respondents. These were aimed at understanding whether they felt there was any inconsistency between communications disseminated at different institutional levels and, in the case of such inconsistencies, which governmental level they believed to be more reliable.

Table 5.3 shows that, despite well-known episodes involving the regions most affected by COVID-19, such as Lombardy and Veneto, the vast majority of respondents from those regions did not report high levels of inconsistency in the way information on restrictions was disseminated at the national, regional, and municipal levels. However, this does not seem to have influenced the opinions of the sample about the different levels of trust they gave to the different governmental levels. When asked "to whom do you give greater credibility among the various levels of government

Table 5.3 Distribution of perceived inconsistency among different levels of government in COVID-19 management

	Very little	Little	Medium	Much	Very much
Restrictions disseminated at the national level are inconsistent with those disseminated at the regional level	**20.7**	**26.5**	**32**	13.3	7.4
Restrictions at the regional level are inconsistent with those at the municipal level	**29.3**	**32.5**	**25.6**	7.2	5.4
Restrictions at the municipal level are inconsistent with those at the national level	**28.7**	**28.5**	**27.4**	9.2	6.1
Restrictions are inconsistent within the neighboring municipalities	**32.6**	**28.7**	**24**	8.8	6

Table 5.4 When you find inconsistencies in the information you receive about restrictions put in place by these levels of government, to whom do you tend to give more credit?

National level	**49.2**
Regional level	**26.6**
Municipal level (my or other municipality)	11.7
None of the three	**12.5**
Total	100

when inconsistency is detected in the management of the situation resulting from COVID-19" (see Table 5.4), the majority of the respondents replied "national" (49.25%) and, to a lesser degree, "regional" (26.6%). There is also a not-insignificant percentage (12.5%) of people who did not consider any of the three levels of government credible sources of information.

In this section, the information processed for each dimension has been synthesized into composite indicators/indices to assess the general trends found in the sample. Given the low relevance of the inconsistency among governmental actors perceived by respondents, this variable has been discarded from the index creation procedure. At the same time, the dimensions of the use of sources, channels, and content and their trustworthiness will be synthesized through the application of principal component analysis.

The first battery of questions to be reduced in size was the one for uses of sources and channels of information (Table 5.5). What emerges from the synthesis of this battery is a breakdown of the information into two components. The first component is entirely characterized by the use of informal sources that form a substratum of shared opinions. Relevant in this index are the use of entertainment programs (television, radio, and newspapers), social media channels, messaging, and the use of requests to acquaintances, relatives, and friends. The second component is the use of authoritative sources aired through information or political debate television programs, scientific websites or publications, civil protection bulletins, governmental and institutional websites, and official online and offline journalistic information.

To make the distribution of the two indices more understandable, the indices were recoded into ordinal variables with three modalities: High, medium, and low. The procedure involved normalization by subtracting from the individual scores obtained on the index the minimum value related to the range of variation and multiplied by one hundred. Subsequently, the percentage distribution was divided into three parts. Figure 5.1 shows that the sample tended to make little use of informal sources. In contrast, the use of authoritative sources is normally distributed across the sample, suggesting a need to look for and use information with a solid and legitimate basis.

Table 5.5 Principal component analysis of the dimensions of the use of sources and channels of information

Rotated component matrix

	Use of informal sources – general opinions	Use of authoritative sources – authority
Television programs of information or political debate	.280	**.406**
Entertainment television programs	**.597**	.152
Radio programs of information or political debate	.336	.317
Entertainment radio programs	**.540**	.214
Scientific web or paper publications	.048	**.558**
Civil protection bulletin	.008	**.739**
Governmental and institutional websites	−.031	**.816**
Official journalistic information websites	.121	**.668**
Opinion or entertainment websites and blogs	**.664**	.198
Social media (e.g., Facebook, Twitter, Instagram)	**.581**	.089
Newspapers and news magazines	.286	**.541**
Entertainment magazines	**.674**	.173
Messaging (e.g., WhatsApp, Telegram)	**.710**	.010
Request to acquaintances whom I consider to be experts	**.536**	.028
Ask relatives/friends	**.718**	−.028

Extraction methods: Principal component analysis. Rotation method: Varimax with Kaiser normalization.

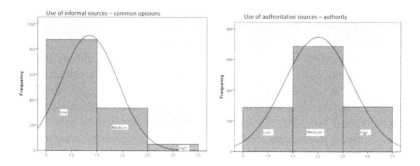

Figure 5.1 Histograms of the distribution of common opinion and authority indexes

The second battery of questions synthesized with the principal component analysis is trust in sources, channels, and content (Table 5.6). The four extracted components refer to (1) trust in official information (print and web newspapers, as well as TV and radio news programs or political debate); (2) trust in institutions (with recorded and live broadcasts of politics, government representatives, and civil protection agencies); (3) trust in science (scientific programs and debate, and recorded or live broadcasts from members of scientific community); and (4) trust in the informal network (as well as in relatives, friends, and acquaintances).

Looking at the distribution of the discretized categorization of these indices in Table 5.6, it is possible to note that the sample tended to place average trust

Table 5.6 Principal component analysis of the dimensions of trust in sources, channels' set of variables

Rotated component matrix

	Trust in official information	Trust in institutions	Trust in science	Trust in informal network
Print newspapers	**.812**	.137	.226	−.042
Web newspapers	**.757**	.111	.207	.077
TV news programs	**.804**	.251	.146	−.094
TV news and political debate programs	**.731**	.266	.089	.076
Radio news	**.831**	.150	.163	.019
Radio programs on information and political debate	**.783**	.161	.102	.118
Political posts and live broadcasts on social networks	.229	**.798**	−.003	.294
Posts and live broadcasts of government representatives	.255	**.858**	.231	−.051
Posts and live broadcasts by civil protection agencies	.269	**.667**	.475	−.118
Scientific programs and debates	.300	.087	**.858**	.065
Posts and direct from members of the scientific community	.157	.235	**.876**	.088
Informal networks (family, friends, acquaintances)	.049	.078	.098	**.958**

Extraction methods: Principal component analysis. Rotation method: Varimax with Kaiser normalization.

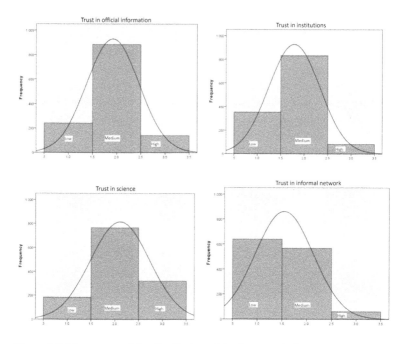

Figure 5.2 Histograms of the distribution of the four trust indexes

in official information sources and science. In contrast, the distribution of trust in institutions and informal networks is negatively asymmetrical, and the sample tended to place less trust in these informal networks. Also, a strong need emerges from the sample to rely on experts to fully understand the pandemic.

Agreement with measures of contrast, containment, and mitigation of the health, social, and economic effects of COVID-19

This section discusses the COVID-19–related information regarding the restriction measures and regulations implemented by institutions in order to face the pandemic. Institutional actions are able to provoke an immediate reaction in the subjects experiencing the pandemic situation, so in our web survey we have focused our attention on the level of agreement with these measures by our respondents. We assume that the level of agreement with governmental regulations might be impacted by people's ontological (in)security. Tables 5.7 to 5.14 show the distributions of the levels of agreement/support with measures related to the closure of schools and

Table 5.7 Agreement level for school and work measures

How much do you agree with the following statements on school and work measures?	Low	Medium	High
Preschools, primary, secondary, high schools, and universities nationwide must remain closed until a vaccine is found.	**57.8**	21.1	21.1
Online learning is the most appropriate way to maintain schools' educational functions during the second phase of the COVID-19 emergency.	38.2	21.6	40.2
Public and private organizations must support teleworking (smart working) during the second phase of the COVID-19 emergency.	14.5	17.3	**68.2**
Companies must respect the ban on economic redundancies related to the COVID-19 crisis during the emergency phase.	12.9	11.8	**75.3**

universities, the implementation of smart working, and the closure of restaurants and museums, just to mention a few. With regard to measures on work and school (Table 5.7), more than half of the sample showed a low propensity to keep schools and universities closed until a vaccine could be found. Online education does not seem to have been for them a compelling element to foster support of generalized closure. At the same time, smart working and a ban on dismissal should be safeguarded and pushed as tools to combat the emergency, according to about 70% of the sample.

Measures related to public and commercial activities (Table 5.8) received mixed answers from the respondents: More than 50% of the sample did not fully agree with maintaining the closure of lodgings, restaurants, and personal care and beauty establishments. However, they did agree with measuring people's temperature at the entrance of public spaces as a preemptive containment measure.

Table 5.9 shows that the majority of the sample agreed with the prohibition of any type of mass event, whether secular or religious, with some hesitation about adhering to restrictive measures when they take place in open public spaces.

When asked about travel restrictions (Table 5.10), the sample, for the most part, agreed on entry and exit restrictions that concern regional, national, and international borders. However, the respondents to our survey were more supportive of restricting the entry of people from China and the USA, who have been more in the media spotlight because of their outbreaks. National solidarity, however, emerges from the limited support given to the phrase, "the southern regions should close their borders to

Table 5.8 Agreement level for public and commercial activities

How much do you agree with the following statements concerning measures on public and commercial activities?	Low	Medium	High
During the second phase of the COVID-19 crisis, e-commerce should be allowed for the same goods that retail shops can sell.	24.9	26.1	49.0
Restaurants and bars should remain closed during the second phase of the crisis.	**58.5**	23.3	18.2
Tourist services (including accommodations, ski and beach services, museums, attractions, etc.) should remain closed during the second phase of the emergency.	**53.8**	23	23.2
Personal care businesses (barbers, hairdressers, beauty centers, and spas) should remain closed during the second phase of the emergency.	**61.2**	20	18.8
People's temperature should be measured when entering public places such as communities, supermarkets, and hospitals.	22.6	15.4	**62.0**

Table 5.9 Agreement level for management of mass gatherings

How much do you agree with the following statements related to measures for the management of mass gatherings?	Low	Medium	High
The need to hold mass events (concerts, sporting events, etc.) must be carefully assessed throughout the country.	11	1.,4	**77.6**
All mass-gathering activities (private parties, spiritual groups, etc.) should be banned, if deemed necessary by the government, at all stages of the COVID-19 crisis.	20.9	13.,4	**65.7**
All crowded public places (not related to daily needs) must be closed, such as large shopping centers, hotels, cinemas, restaurants, and so on, if deemed necessary by the government, in all phases of the COVID-19 crisis.	32	21.7	46.3
If deemed necessary by the government, religious services should be banned in all phases of the COVID-19 crisis.	27.4	16.5	**56.1**

Table 5.10 Agreement level on travel restrictions

How much do you agree with the following statements regarding measures for travel restrictions to contain COVID-19?	Low	Medium	High
It is necessary to control traffic and private travel of people in and out of regions.	27.9	20.5	**51.6**
It is necessary to control national public transport traffic (trains, ships, planes, buses, etc.).	18.8	19.9	**61.3**
Limit the entry into Italy of people coming from China's risk areas.	24.3	20.3	**55.4**
Limit the entry into Italy of people coming from the USA.	23.8	20.6	**55.6**
Limit the entry into Italy of people coming from Europe.	32.7	22.7	**44.6**
Limit the entry into Italy of people from Africa.	31.2	23.2	**45.6**
The southern regions should close their borders to people from the North's most affected regions (Lombardy, Piedmont, Veneto, Emilia).	41.4	20.6	38.0

Table 5.11 Agreement level for home isolation and quarantine measures

How much do you agree with the following statements regarding home isolation and quarantine measures?	Low	Medium	High
The government should call on citizens to control exiting their homes unless it is an emergency.	33.5	23.5	43.0
Patients with mild symptoms of COVID-19 should be isolated and treated at home and should only contact their own doctors.	18.0	18.0	**64.0**
All people who have returned to Italy from a country where the outbreak has occurred should undergo 14 days of home isolation and contact their doctor if symptoms are suspected.	9.3	10.5	**80.2**
People who have been in contact with a confirmed case of COVID-19 should undergo home isolation for 14 days and contact their doctor if symptoms are suspected.	6.1	8.8	**85.1**

people from the most affected regions in the North (Lombardy, Piedmont, Veneto, and Emilia Romagna)," showing that the risk was more accepted within national borders.

The table's analysis related to measures on home isolation and quarantine (Table 5.11) shows that the sample was not decided on placing the responsibility for control and compliance directly on the citizen. At the same time,

Table 5.12 Agreement level for sanctions spread during the emergency

How much do you agree with the following statements regarding the sanctions imposed in order to manage the COVID-19 emergency?	Low	Medium	High
Violators of measures to contain the epidemic (such as wearing a mask, social distancing, limited exits only in case of emergencies or to reach workplaces) are subject to a fine (from €400 to €3,000).	22	15.7	**62.3**
Failure to comply with quarantine measures by anyone who has tested positive for COVID-19 carries criminal penalties: imprisonment from 3 to 18 months and a fine of between €500 and €5,000, with no possibility of ablation.	13.8	10.2	**76.0**
Violating the quarantine and having contracted the virus, leaving the house, and spreading the disease may lead to prosecution for serious offenses (epidemic, murder, injury), punishable by severe penalties of up to life imprisonment.	15.3	11.4	**73.3**
Self-certification (printed or handwritten) is required to leave the house in case of legitimate need during the COVID-19 emergency.	34.7	19.7	45.6

the vast majority agreed with all restrictions and protocols disseminated on quarantine, isolation, and disease detection prophylaxis.

With regard to sanctions (Table 5.12), the majority of the sample supported not only the use but also the congruity of the punitive measures adopted, with the sole exception of the use of self-certification, perhaps because it was repeatedly subjected to changes and variations, leading those confronted with it to be somewhat confused about its use and usefulness.

The vast majority of the sample considered the income support measures necessary to support families with children, the elderly, the disabled, and workers unable to return to work or resume their activities (Table 5.13).

With regard to the measures designed to ensure health (Table 5.14), while the sample generally considered it mandatory to wear a mask and to sanitize themselves, their objects, and their environments, the same degree of agreement was not reached with regard to the imposition of the restriction on going out in order to reduce the opportunities for contagion and maintaining a safe distance even from relatives. This latter result is not insignificant, since it shows that standardized and mandatory behavior (masks and sanitization) were accepted to a greater extent, and that behavior involving personal restrictions, sometimes self-imposed (interpersonal distance and going out), were much less accepted, revealing the extent to which

Table 5.13 Agreement level for income support measures

How much do you agree with the following statements regarding income support measures?	Low	Medium	High
It is necessary to consider extraordinary support measures for households with children, to cope with the emergency caused by the outbreak of COVID-19 at least until Phase 3, such as babysitting bonus, parental leave, etc.	5.1	14.2	**80.7**
It is necessary to think about extraordinary support measures for families with elderly or disabled people, to cope with the emergency caused by the outbreak of COVID-19 at least until Phase 3, such as an increase in paid leave days under Law 104/92.	4.1	11.4	**84.5**
It is necessary to think about extraordinary income support measures for workers unable to return to work, to deal with the emergency caused by the outbreak of the COVID-19 epidemic, at least until Phase 3, such as the redundancy fund and compensation for VAT payments.	3.3	9.4	**87.4**

Table 5.14 Agreement level for measures to ensure health

How much do you agree with the following statements regarding measures to ensure health?	Low	Medium	High
Wearing masks should be compulsory for everyone.	16.1	10.1	**73.8**
Stay indoors to avoid creating aggregations or opportunities for contagion.	34.6	21.8	43.6
Wash hands, ventilate, and disinfect surfaces with which you come into frequent contact.	6.2	7.9	**85.9**
I must keep a distance of one meter between myself and others when I go out, even if they are relatives living together.	34.3	16.6	49.1

containment measures involving isolation from the spaces and people of one's pre-pandemic daily life take a greater toll on the individual than those involving masks and sanitizing.

A dimensional reduction analysis based on principal component analysis (PCA) was carried out to synthesize all the information from the variables just described. The variables that were not relevant in the sample (as described above) were removed. The PCA was conducted on 24 variables out of 35 used to detect the different levels of agreement with measures on different dimensions (school, work, welfare, health, sanctions, etc.).

The main component analysis returned an index that encompasses 47% of the total variance in the data and expresses the level of agreement with regulatory measures. As seen, however, since the level of agreement shown by the sample on many of the dimensions was very high, in the procedure of discretizing the values of the index, instead of resorting to tertiles, the values were dichotomized by joining the low and medium classes and leaving the high class alone.

The resulting distribution is represented in Figure 5.3 and shows the evident prevalence of a high degree of agreement to regulatory measures in the sample.

Table 5.15 shows a greater level of agreement with the measures by women as compared to men. There was less agreement in the more affected regions and more agreement in the less affected ones. This may owe to the fact that the former were already affected by the severity of the virus (and, therefore, already active in facing the virus, perhaps accepting the restrictive measures less willingly); meanwhile, the latter probably feared COVID's advance by looking at what was happening in the North of the country. On the other hand, differences based on age and socioeconomic status of the sample are not significant.

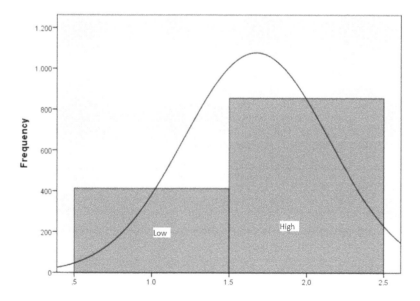

Figure 5.3 Histograms of the distribution of the dichotomized index

Table 5.15 Measure agreement for principal sociodemographic variables

		Low	*High*
Gender	Female	45.1%	**54.3%**
	Male	**54.9%**	45.7%
Macro-area of	Northwest	**26.4%**	26.8%
residence	Northeast	**26.2%**	16.0%
	Central	16.5%	**21.5%**
	South	20.3%	**24.4%**
	Islands	10.7%	11.2%
Age group	Youth (18–28)	16.5%	18.6%
	Young adults (29–39)	31.0%	29.5%
	Adults (40–59)	41.4%	38.3%
	Seniors (60+)	11.1%	13.5%
Socioeconomic	Low	4.6%	3.2%
status	Low-intermediate	36.6%	31.7%
	Medium	16.0%	20.8%
	Upper-intermediate	40.9%	40.2%
	High	1.9%	4.2%
	Total	100%	100%

This whole system of direct and indirect interconnections, whether proportional or not, can be further explored if the information produced through the processing of the indices created in this chapter is brought back into the ontology schema elaborated in the previous chapter, trying to give further specifications to the three emerging ontological profiles.

The public sphere effect on the COVID ontological profile framework

In this final section, we attempt to answer the following research question: What are the interactions of the ontological (in)security dimensions from the public sphere and private sphere?

To answer this question, all the indices created were put under a multiple correspondence analysis as active modalities (i.e., competing in the determination of the synthesis sought factors). The other variables with an active role were also the indices related to the private sphere influences: indices of fear, knowledge, experience proximity, geographical proximity, reprimand to strangers or family members, perception of the usefulness of one's behavior, the importance of religious belief, political orientation, cultural worldview, level of adherence to the adoption of protective behaviors, and level of adherence to the respect of the restrictive measures. The variable related to reliability conferred to certain governmental levels (nation,

Table 5.16 Measure agreement for detected indices

		Low	High
Fear	Low	**44.39%**	28.41%
	Medium	46.17%	40.66%
	High	9.44%	30.93%
Knowledge	Low	**45.52%**	34.74%
	Medium	33.66%	**44.80%**
	High	20.82%	20.47%
Experience proximity	Low	49.88%	48.59%
	Medium	35.84%	35.48%
	High	14.29%	15.93%
Geographical proximity	Low	30.99%	**35.67%**
	Moderate-low	16.46%	21.52%
	Moderate-high	**26.15%**	16.02%
	High	26.39%	26.78%
Level of adherence to the adoption of protective behaviors	Low	1.94%	3.51%
	Medium	19.85%	9.60%
	High	78.21%	**86.89%**
Level of adherence to compliance with restrictive measures	Low	13.,56%	**26.11%**
	Medium	65.38%	67.10%
	High	**21.07%**	6.79%
Egalitarianism	Low	8.25%	5.15%
	Medium	**38.35%**	29.71%
	High	53.40%	**65.15%**
Hierarchism	Low	**54.85%**	21.31%
	Medium	38.83%	**52.34%**
	High	6.31%	26.35%
Fatalism	Low	36.32%	**41.17%**
	Medium	**57.63%**	49.36%
	High	6.05%	9.47%
Individualism	Low	48.30%	43.27%
	Medium	42.48%	47.02%
	High	9.22%	9.71%
Common opinion	Low	67.23%	70.18%
	Medium	29.37%	25.38%
	High	3.40%	4.44%
Authority	Low	**33.74%**	17.33%
	Medium	55.10%	53.86%
	High	11.17%	**28.81%**
Trust in official information	Low	**26.39%**	15.56%
	Medium	67.80%	70.76%
	High	5.81%	**13.68%**
Trust in institutions	Low	**37.14%**	23.30%
	Medium	59.47%	68.74%
	High	3.40%	**7.96%**
Trust in science	Low	**17.96%**	12.65%
	Medium	64.08%	58.43%
	High	17.96%	**28.92%**
Trust in informal network	Low	38.35%	**56.32%**
	Medium	**54.85%**	39.93%
	High	6.80%	3.75%
	Total	100%	100%

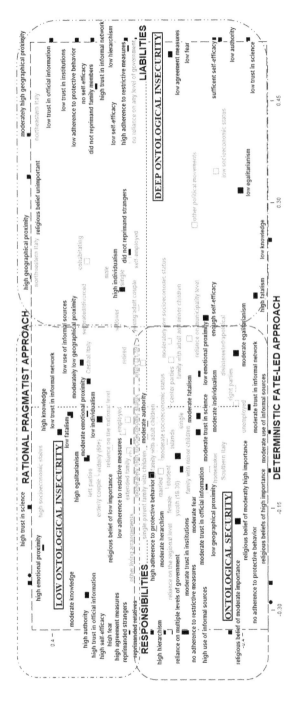

Figure 5.4 Factorial plan generated through multiple correspondence analysis on private and public sphere COVID's influence factors, with the projection illustrative of sociodemographic characteristics of the sample

region, or municipality) because of its unbalanced distribution was given a role of explanatory variable (i.e., useful in better defining the synthetic factors extracted), together with the main sociodemographic characteristics of the sample (sex, age group, macro-areas of residence, socioeconomic status, occupational status, and family type). All these features were projected onto the factorial plan (see Figure 5.4) to test whether the ontological groups that were traced hold up to the inclusion of new dimensions and add interesting connotative elements to the emerging profiles.

The two summary dimensions, latent variables, components, or factors, extracted through the MCA, retain the same meaning as extracted in the MCA discussed in Chapter 4 and are underlined by the addition of influence elements from the wider public sphere. The total inertia synthesized by these two components decreases due to the inclusion of additional information contained in the added variables and indices and goes from 55% to 53.5%. They can be described as follows:

• The first factor (horizontal axis, with 30.2% of inertia explained): *Responsibilities vs. Liabilities.* With this factor, an opposition emerges between a proactive behavioral dimension, in which the belief and value system plays a key role in inducing the adoption of protective behavior and a high agreement with the control, containment, and protective governmental measures. This is beyond the fact that it becomes the subject of regulatory containment measures, with a high use of formal and informal sources of information and more trustworthiness conferred on official information and institutions (left-hand side). There is also a passive behavioral dimension in which beliefs and values lose their significance, and the lack of perception of the importance of doing one's part in combating the pandemic leads to formal adherence to restrictive measures without, however, adherence to the responsible adoption of protective behavior. This is reflected in a low level of agreement with governmental measures, and a lack of trust in institutions and official information, in favor of a greater trust conferred on an informal network (right-hand side of the plan);

• The second factor (vertical axis, with 24.3% of inertia explained): *Rational Pragmatist Approach vs. Deterministic Fate-Led Approach.* With this factor, the opposition system is between types of approach to risk perception. In the upper part of the plan, a rational and pragmatic component prevails, based on moderately high levels of knowledge, awareness, and perception of geographical and emotional proximity, together with a collectivist and egalitarian cultural approach and trust conferred on science. On the other hand, in the lower part of the plan, there is little perceived knowledge of how to protect oneself, little knowledge of the danger of proximity, and a value system tending toward fatalism that considers the events of the world to be governed by a predetermined destiny, already established and, therefore, subject to social determinism. This is coupled with a high importance conferred on religious belief and only moderate trust in informal networks.

The two factors were then cross-referenced to create a factorial space of attributes to project the position of all the modes of the active and illustrative variables used in the analysis. Figure 5.4 shows this plan. A cluster analysis was developed to further synthesize the characteristics investigated into homogeneous groups. By applying semi-hierarchical cluster analysis, three groups were identified, summarizing respectively 29%, 43%, and 28% of the individuals' variance in the sample.

These groups retain all previous characteristics, to which new details can be added. In the medium ontological insecurity group, in the upper-left quadrant of Figure 5.4, and born from the intersection of an active, asset-led behavior and a rational, pragmatic approach to risk perception, we can notice that by broadening the system of influences to include public ones, some characteristics in the group vary. Structurally, this group is mainly made up of elderly (over 60 years old) people, who are workers or retired people; they live as couples, are married, or take care of single-parent families as widowers, with a socioeconomic status tending to high and a political orientation that moves through the left. Here a moderately high perception of the virus' emotional proximity is prevalent, as is moderate knowledge and awareness; there is a high perception of fear, but also of the usefulness of one's own behavior in facing the virus's advance, so much so that this group presents a solid adherence to the adoption of protective behavior to the detriment of low adherence to restrictive measures merely because these are regulated. These determinants of the private sphere are accompanied by an openness to the public sphere, characterized by a high degree of recourse to authoritative sources of information, conferring a high degree of trust on science and official information, which leads to a good combination of knowledge, awareness, and updating with respect to the situation that generates a higher level of agreement than the containment, control, and protection measures of a governmental nature that were designed to stem the advance of the virus. This is further underlined by giving greater reliance to the national level of government.

Moreover, as just highlighted in Chapter 4, the space of ontological security is governed by a low individualism and a high egalitarian conception, all elements found in the implementation of proactive behavior and also expressed in the action of reprimanding strangers or family members if they are caught in the act of contravening the rules that currently dominate the scenario of securing against the possibility of contracting COVID-19. This is a group of people who are extremely responsible for themselves and for others, both near and far, and for whom medium ontological insecurity becomes a survival mechanism in which risk is accepted, understood, and processed. This group chooses to adopt moral behaviors that are not intended to be regulated or sanctioned but should become part of the responsibility of each individual to society as a whole. This is reinforced by reliance on official sources, on the power of science and knowledge to generate specific responses to risk, and on a reliance on ad hoc measures by the national government.

Medium-high ontological insecurity, the group on the right side of the plan, is characterized by a passive behavior orientation. A rational and more fatalistic component prevails in a mixed way. This group is primarily made up of men from northern or central Italy, tending to be adults, single or cohabiting, and self-employed, who identify their political orientation with other movements or formations and have a high or moderate socioeconomic status. Compared to the previous group, they do not have families with children, so the component of caregivers is missing, and the propensity for individualism is strong, probably driven by professional motivations that make the components of this group capable of accepting the risk even when they know how dangerous it is; otherwise, it would become unmanageable to live with this type of concern. In this group, characteristics such as moderately high geographical proximity of the virus, high awareness, and self-perception with regard to knowing how to protect oneself or to the ability of others to protect themselves from the virus coexist with a decline in the importance of values such as religion and the influence of the authoritarian component. This latter behavior is typical of this group, which declared a low adoption of protective behaviors and a high level of adherence to restrictive measures because they are regulated, thus underlining the characteristic liability component of this group. In this group, in fact, the level of agreement with governmental control and with containment and protection measures was very low, if not absent; little trust was placed in official information, institutions, and science. At all levels of government, this group tended to reject the authority of these sources and entrust it, instead, to informal networks and information channels. They still showed little concern, despite the high geographical proximity and close knowledge of the phenomenon, which seems to delimit a mechanism that leads to the acceptance of a situation of risk, almost ignoring it (along with the authoritative source), and remaining therefore passive. This is even more evident when it comes to the group's adherence to a high level of fatalism that guides their lives. They tended to ignore the capacity for self-determination and individual responsibility, highlighting the lack of credibility of the actors involved in the management of the emergency.

Ontological security is the last group of detected frames fitting with the pandemic emergency. It is positioned in the lower-left quadrant of

Figure 5.4, in the intersection of a deterministic, fate-driven approach to risk perception, with a prevalence of proactive, responsibility-led behavior. This group is more likely to include the youngest members of the sample, as well as adults; it is made up of singles but also of families with children. The members of this group are generally less wealthy, because of their moderate socioeconomic status, even tending to the low end. Its members are sometimes still students or unemployed or housewives, who go from showing a tendency to disinterest in politics to a timid positioning to the right. They reside, for the most part, in the South or on the islands; and while, on the one hand, this may justify a more distant perception of the risk (given that, in the initial phase of diffusion of COVID-19, Italy saw the northern areas of the country most affected), on the other hand, it does not justify their low level of concern – all the more so given the fact that many of the respondents in this group are also parents and, therefore, the caregiver component can be decisive for them. This shows, in a way that leaves little room for doubt, how much effect a reaction aimed at medium high ontological (in)security as a mechanism of conscious acceptance of risk is mostly the professional component. Meanwhile, to effect the construction of a system of ontological security in which one does not get to understand well the situational context of risk within which one is moving, this affects in a more decisive way both family composition and socioeconomic status. This is also connected to relevant educational poverty and open scenarios of contextual possibility, just because the part of the country in which this type of group is concentrated is the South, notoriously affected by various gaps and systems of inequality. On the side of private sphere influences on COVID perceptions, poor knowledge prevails in this group, along with low awareness and low emotional and geographic proximity to the diffusion of the virus. People in this group positioned themselves on average values. For instance, people presented moderate traits of hierarchism, egalitarianism, individualism, and fatalism, in conjunction with the influence of religion ranging from a low to a high level, while having a sense of enough self-efficacy. On the side of public sphere influences, instead, this group chose to stay in the middle, placing media trust in science, official information, and informal networks, and yet relying on institutions (mostly local ones recognized in regional and municipal levels of government) and on the use of informal sources to retrieve information about the virus. This group did not support protective measures and did not adopt protective behaviors to safeguard themselves from the pandemic threat, by which it seems they did not feel threatened – perhaps because the surrounding physical environment was not threatening. The southern part of Italy was not really affected during the first and second phases of response and recovery. Also, the people in this group primarily relied on unofficial information that, compared to political and scientific debates, contained less technicalities, alarmism, and clarifications. Thus, this group seems to be characterized by a complete lack (or at least by a very low) of risk perception, which puts them in a sort of risk denial status, in which the adoption of protective behaviors may be perceived as unnecessary.

This chapter's main finding shows how the private influence component and the public influence component tend to complement each other and define increasingly stable ontological profiles in dealing with pandemic risk. However, if these three are the profiles our sample shows in adopting or not adopting protective behaviors, what does this sample expect in the near future? What policy and practical governance lessons have been learned in the management of the second phase of the COVID-19 crisis in Italy? And why might a reflection on these matters be useful with respect to future scenarios? These general closing questions will be answered in the final chapter.

References

Cinelli, M., Quattrociocchi, W., Galeazzi, A., Valensise, C. M., Brugnoli, E., Schmidt, A. L., & Scala, A. (2020). The COVID-19 social media infodemic. *Scientific Reports, 10*, 16598. https://doi.org/10.1038/s41598-020-73510-5

De Falco, C. C., Punziano, G., & Trezza, D. (2021a). A mixed content analysis design in the study of the Italian perception of the COVID-19 on Twitter. *Athens Journal of Social Sciences*, 8(3), 191–210. https://doi.org/10.30958/ajss.8-3-3

De Falco, C. C., Punziano, G., & Trezza, D. (2021b). Digital mixed content analysis perspective for the study of digital platform social data: An application on the analysis of the COVID-19 risk perception in the Italian Twittersphere. *Journal of Mixed Methods Research*. Forthcoming.

Lindell, M. K., & Perry, R. W. (2012). The protective action decision model: Theoretical modifications and additional evidence. *Risk Analysis: An International Journal*, 32(4), 616–632.

Mesquita, C. T., Oliveira, A., Seixas, F. L., & Paes, A. (2020). Infodemia, fake news and medicine: Science and the quest for truth. *International Journal of Cardiovascular Sciences*, 33(3), 203–205.

Reynolds, B., & Seeger, M. W. (2005). Crisis and emergency risk communication as an integrative model. *Journal of Health Communication, 10*(1), 43–55.

Ruiu, M. L. (2020). Mismanagement of COVID-19: Lessons learned from Italy. *Journal of Risk Research, 23*(7–8), 1007–1020. https://doi.org/10.1080/136698 77.2020.1758755

Wendling, C., Radisch, J., & Jacobzone, S. (2013). *The use of social media in risk and crisis communication*. [OECD Working Papers on Public Governance, No. 24]. https://doi.org/10.1787/5k3v01fskp9s-en

6   The world after COVID-19

Introduction

This chapter discusses the vision of a COVID-19–free future in Italy at both individual and collective levels. From the individual perspective, the study shows whether the respondents believe their private and social lives will change once the pandemic crisis ends. At the societal level, the COVID-19 turmoil, even though it has created severe consequences worldwide, has also provided a significant opportunity to reshape policy frameworks toward a more sustainable economy and society. This means working at the level of desires and expectations, without losing grip on the emotions, experiences, and perceptions that accompanied our sample in the tailspin of the first pandemic wave in Italy.

The chapter is divided into four parts, discussing (1) how Italians see themselves in the future; (2) the effects that COVID-19 management have had on Italians' perceptions of the trustworthiness and credibility of governmental and health organizations; (3) analysis of the strengths and weaknesses of the management of the crisis according to our respondents; and, finally, (4) suggestions for improving policy, practice, and social science research.

The first part of this chapter focuses on the fears and hopes of the respondents regarding the near and distant future. The COVID-19 outbreak has sowed the seeds of change in many facets of daily life: Work, family, social and love relationships, leisure, and many other social spheres are facing profound changes that could be enduring, or could revert to their traditional forms as soon as life goes back to some sort of normality. However, it appears that there are many fields in which the measures implemented to address the COVID-19 crisis could have created long-lasting effects. For example, our results show that Italian respondents are now getting familiar with the idea that remote working is a viable and effective solution; this is a

DOI: 10.4324/9781003187752-6

surprising result, considering that the Italian job system is generally slow in implementing innovations in working conditions.

The second part of the chapter focuses on how the COVID-19 crisis boosted or lowered people's trust in the national government and institutions by trying to understand the reasons for disappointments and comfort zones that risk management directly generates in terms of risk experiences.

The third part of the chapter will be dedicated to analyzing the strengths and weaknesses in the management and governance of the crisis, starting from the individual reflections provided by our sample and trying to systematize this recognized knowledge into policy suggestions and concrete actions to improve people's resilience, with a better policy direction that looks to the future – with or without the virus.

Lastly, a distinct final section discusses some critical and epistemological reflections on the future of social research after the COVID-19 crisis, highlighting how the choice of the web survey method is a direct consequence of both our research object and its timing: A survey during a pandemic would not have been possible if not conducted via the web, just as such a rich and in-depth questionnaire would never have found responses if the general lockdown had not forced the respondents to "stay at home," to have a great deal of time available (due to the limitation of all daily activities), but above all to have developed that attitude of being "always online," which gives the feeling of being less alone in a period of forced isolation.

Reflections on the post-COVID-19 future: the fears and hopes of the respondents

Fear and hope for the future are the feelings that arise first and foremost in individuals when they are confronted with an unknown enemy of whose management they are not certain. It is for this reason that, in this section, the main issue is the analysis of a question posed to the sample: "How do you imagine your future after the end of COVID-19?"

Results show that more than half of the sample is optimistic about the future, since they think that there will be no change after the end of the COVID-19 crisis (Table 6.1). By contrast, 37.4% ("worsened" + "much worse") believe things will get worse. Only 10% of the sample is extremely optimistic, as they think that their lives will be improved at the end of the pandemic.

Reinforcing this result is another that sums up the feelings respondents had about the possibility of being infected. The sample was asked "how does the possibility of falling ill with COVID-19 make you feel?" This was a closed question with the possibility of specifying as-yet uncategorized emotions.

Table 6.1 How do you imagine your future after the end of the COVID-19 emergency?

Response	%
Much improved	3.2
Improved	7.0
The same	**52.4**
Worsened	**26.9**
Much worse	**10.5**
Total	100.0

Table 6.2 The possibility of falling ill with COVID-19 makes you feel . . .

Response	%
Alert/attentive/concerned for myself and others	**47.6**
Nervous/angry/irritated/anxious/confused	**18.1**
Scared	**16.3**
Indifferent/not afraid/fatalist	9.5
Depressed/bored	5.0
Optimistic/energetic	3.5
Total	100.0

Results show that the prevalent answers concentrate on feelings of alertness, attention, and concern for oneself and for others (47.6%). Stress, accompanied by feelings of being nervous, angry, irritated, anxious, or confused, characterized 18.1% of the sample. Fear is a condition reported by 16.3% of the respondents, while indifference and fatalism (9.5%) and depression and boredom were less reported (5.0%).

When asked "when do you expect the crisis brought about by COVID-19 to end and life to return to normal?" (Table 6.3), two-thirds of the respondents believed this would happen at some point in 2021 or later, showing an awareness of the need for the health and socioeconomic situations to evolve; meanwhile, 27.6% believed that it may end by 2020, showing an optimism that, in the short term, has been shown to be completely unwarranted by the facts.

The reasons for being optimistic, despite the alert and the feeling that the virus may have a negative influence on the near future, are diverse and varied. We have highlighted the most significant and representative answers for each position.

Scrolling through the list of motivations provided by the respondents with regard to how they imagine their future after the COVID-19 crisis is

Table 6.3 When do you expect the crisis caused by COVID-19 to end and life to return to normal?

Response	%
By 2020	27.6
At some point in 2021 or later	**66.8**
Never	5.6
Total	100.0

over, there are some recurrent patterns in the answers that could be categorized as follows:

- *Concerns about the economic crisis.* According to the respondents, the prolonged restrictive and containment measures will have a potentially significant impact on the country's economy. They complain about the changes in work experiences, many of which have been or will be interrupted as a result of COVID-19. They complain about the lack of measures to support the role of women workers, as well as the education and employment of the youngest. And there is also a fear of a potential rise in crime and illegal behavior.
- *Concerns about the social and psychological effects of the crisis.* There is fear that, as an effect of isolation, people will develop a sort of inability to manage various forms of socialization. Respondents said, "I'm afraid to go out and do the normal things I used to do. I can't imagine going on a holiday, a plane, or a boat trip. I can't imagine going to the cinema or to the theatre, to a restaurant or to other places where people gather." This climate of uncertainty is hindering the ability to plan in the short, as well as in the long, term. Other consequences are (1) the fear of the other, due to lack of reasoning and responsibility (as seen in statements like, "I don't think people have really learned anything from this pandemic and I firmly believe that people will go back to behaving as before if not worse"; and "human beings forget the fear and make the same mistakes over and over again"); as well as (2) the collapse of trust in the authorities responsible for managing the pandemic.
- *Concerns about digitalization.* All processes of social, educational, work, administrative, and relational life have been experienced online during the pandemic. According to some respondents, this will lead to the collapse of traditional ways of living, in favor of a completely dematerialized scenario in which everyone can be everywhere and do everything but be deprived of presence and co-presence in every activity: "Normality is a very distant idea. I am afraid that this quarantine

has made people even more attached to technology and increasingly distant from reality. I hope that smart working or distance learning will not become the norm."

• *Concerns about the inability to re-enter the known mechanisms and normal processes of development in different dimensions.* Respondents expressed concerns about how life will look after the pandemic: For students in their course of study; for the unemployed in their ability to find new employment or their place in a world that is now completely inhibited to them; for families to be able to disentangle routines; for children to grow up among others, and in the world with all its beauties; for companies to reopen; for sports, art, and entertainment to regain their centrality; and for mobility and travel to once again bring gratification to the human experience. They said, "economically it will be a disaster, many businesses will be forced to close, some are already closing, not to mention the many people who are no longer here and who will have no future because of this virus."

• *Concerns about the widening gaps between systems of digital, gender, generational, and status inequalities.* Respondents said, "the situation defined by this pandemic has only accelerated and increased what were, and are, inequalities in our society"; "I think there will be a very strong economic and social crisis, because people will not have a chance to get ahead, and eventually a 'social war' will break out between the poor."

There are also other motivations that emerged from the answers of the respondents. These motivations were few, but significant, when it comes to the meaning of the pandemic. The pandemic has been seen by our respondents as:

• An opportunity to break the barriers of physicality, finding new mechanisms of gain;
• A driving force to take back the reins of one's life, one's freedom, and one's capacity for arbitrariness, freeing oneself from pre-imposed logics at every level (social, familial, among peers) in order to assert oneself or find meaning, function, and one's place in the world;
• An engine of community, of solidarity, of re-interest for others and for the community to be cared for, but also as a drive to appreciate the little things, even everyday things, that have been lacking due to restrictions;
• An aid to regaining possession of time and of one's ability to reflect, and to give value to things, people, spaces, and experiences; and
• A means of acquiring awareness: "People have become aware and conscious that science and medicine cannot always guarantee health, so I believe that the most intelligent people will, in the immediate future, adopt lifestyles designed to improve their health and their environment."

Still others expressed a general aspiration that everything would slowly return to how it was before, to what is perceived and defined as "normality," recovering the frame of *ontological security* discussed in Chapters 4 and 5.

Participants in the study tended to manage their expectations, anticipating a gradual return and not rushing the reopening, in order to avoid the possibility of both new waves of infection and new frustrations. Their hope is underscored by drawing on the lessons of history, which have shown us that great catastrophes, wars, and economic depressions have been followed by formidable periods of economic expansion and growth, so that not only will we emerge from this crisis, but we will do so stronger than before. We see here a certain organicist component, as a universal law that guides nature, society, and humanity, always re-establishing the natural flow of things.

There are still, however, those who believe that "your future does not depend on COVID but on your choices and the opportunities that will present themselves." This shows an aversion to fatalism and a focus on individuals as capable of determining their fate.

Some of our respondents expressed a belief that the pandemic will lead to momentary upheavals that will upset the order of things temporarily, but without major imprints on the future. One respondent said that

> the virus, contrary to what many are saying, will not create a better society, based on mutual aid or inclusiveness, nor will there be more social justice. On the contrary, I believe there will be a polarization of mutual positions and roles within society: Those who were in a strong position, compared to others before the pandemic, will emerge even stronger, because they will have benefited from the momentary weakness of others; conversely, situations of economic uncertainty will only be exacerbated. The effects will be visible in the long term in terms of the possibility of finding work or (for many entrepreneurial categories) of returning to former earnings.

By contrast, others believe that the virus will act as a social leveler, as a tool that democratically does not look at prestige or wealth: "Advantages and disadvantages will be balanced: Less income but more time with the family. Less travel but less stress."

There were also very precise and far-sighted analytical considerations regarding what will happen in the immediate future, revealing a profound analytical capacity that characterizes some of the subjects in our sample:

- "The main thing is that I think most people still do not realize that they have to change their lifestyle and pay attention, at least for several months. But today, I understand that with Phase 2 many people have taken it all as a free-for-all and do not pay attention as they should – people running around without masks, crowds, etc. . . . For me and my family, it's still too early; we only go out (as we always did during the most acute phase) every ten days for shopping, and since we've gone back to work our trips are only home–work trips. We don't visit relatives and friends. . . . My children have been at home since March 3rd (school closures). Here, I find that the government has made many mistakes that could have been avoided, such as putting a limit of € on spending and that, in the same household, only one member could go out to shop. On the other hand, on the four occasions that I have gone out shopping, I have seen people come in and leave with just one small bag of bread. . . . Everyone knows that. Even now, in my opinion, it is very wrong to give the 'go-ahead' to go and see relatives and friends. . . . It is not possible. . . . This means that at least four or five people at a time will find themselves in a single house. . . . And in this way, the contagions will start to increase again. . . . I would have left the restrictions until July 31st . . . leaving only . . . shopping/work/home, and that's it. To me, that's the only way to keep the numbers from going up again. And the walks then . . . only with family members. . . . Unfortunately, this is not the case!!!! There are already crowds . . . too many people in other people's homes, etc. So, we'll never get out of this!"

- "I trust in the reflective stimulus on global societies. Today it was COVID, tomorrow other environmental pressures that will expose the limits of this socioeconomic system. So, I trust that I will be able to interact with people who have reached this maturity, which I have been developing for a long time, especially for work reasons, dealing with sustainability."
- "A few months are not enough to direct a substantial change of life. Our established habits of pseudo-wellbeing and consumerism are too widespread, our perception of comfort too institutionalized."
- "The quarantine has highlighted many things in the lives of individuals, and this contributes to reviewing behaviors assumed out of inertia. We now have the opportunity to establish a starting point from which to build a new future that is more closely under one's control, avoiding making mistakes also in interpersonal relationships."
- "As history teaches us, a crisis is also a moment of change. However, it is necessary that this opportunity is well perceived and that it is accompanied not only by the will of individuals to rethink themselves in new terms, but first and foremost by the national government, which has the task of processing failures rather than successes and redefining a horizon for the Italian system, finally adapted to the challenges that other countries are showing they know how to take up and put to good use.

Economically, we will face a recession. Europe faces a test of resilience with uncertain results, social anger that is difficult to manage, problems due to the lack of information on the part of much of the population, due to bias and the practice of informing oneself on social networks and questioning science."

• "I do not imagine a return to normality. I strongly desire that human beings acquire awareness about the protection of the environment, nature, and the planet that hosts us. I imagine a less selfish, less consumerist, and less capitalist world, not governed by political power or big lobbies. I am optimistic and cultivate positive thoughts. Because thought creates matter."

Not to be underestimated is the fact that about 20% of the respondents were unable to formalize the reasons why they perceived changes linked to the virus and its impacts, showing an attitude somewhere between confused, absent, and disinterested in the facts of the world around it, or rather approaching that frame of ontological security whereby one lives in a space in which one shuns what is evident, unable to rationalize it in the space of one's existence.

Reflections on the post-COVID future: risk management impacts risk experiences

This section focuses on how the COVID-19 crisis boosted or lowered people's trust in the national government and institutions by trying to understand the reasons for disappointments and comfort zones that risk management directly generates in terms of risk experiences. The positive judgment of Italians regarding the management of the pandemic might surprise us when we consider that Italians have little confidence in the ability of both national and regional levels of government to manage a second pandemic wave (see Chapter 4). However, among the respondents who perceived an inconsistency between the national, regional, and municipal levels of government, there was a tendency to give more credibility to the national level, followed by the regional; there was, though, a considerable percentage of people who did not consider any of the levels more credible (see Chapter 5). Regarding their trust in institutions, the sample replied by giving an average level tending toward low.

Respondents were asked to rate on a five-point Likert scale, from "very poor" to "very good," the following question: "How would you describe the institutional management of COVID-19 in your country?" More than half (63%) of the sample rated the management of COVID-19 between average and good (see Table 6.4). What motivates this perception? To better

Table 6.4 How would you describe the institutional management of COVID-19 in your country?

Response	%
Poor	11.2
Fair	16.,2
Average	**32.3**
Good	**31.1**
Excellent	9.2
Total	100.0

understand this result, the sample was asked an open-ended question aimed at investigating the deep-seated motivations that move the sample in a moment of extreme difficulty to still show positivity.

Positive opinions cover different spheres of institutional and public action. Many respondents stated that those involved in regional and national governments have done their best to contain the spread of the virus, and that front-line personnel (such as medical and health personnel) have made heroic gestures. Yet, our respondents also justified the choices of politicians who, confused by the uncertainties emerging, did not shirk their organic responsibilities, even when it came to making unpopular but necessary decisions. (Think of all the government measures restricting commercial activities and daily life that were imposed on the population, leaving no room for repetition, but operating a system of compensation and social support measures that could stem the deleterious effects of these measures.) The Italian management of COVID-19 also triggered a nationalistic spirit that made Italians proud of the rapid, efficient, and effective management of the crisis, judged as being comparable to that of some of the stronger European countries, such as Germany, Sweden, and Great Britain. However, respondents also acknowledged that the Italian response model still needs improvement. The feeling of national pride arises, in part, from the fact that Italy was one of the first movers on very uncertain terrain. In the words of one of the respondents, the source of the national pride is "for having been a test country in the institutional situation we have been in for years. I think it has been dealt with and managed as a complex situation in all its facets."

There is also strong support for Prime Minister Giuseppe Conte, who has become a guide in the panoply of measures and feelings that have pervaded the population. What is bragged about is the responsibility of the governmental systems involved, the fact that they put their noses to the grindstone and did their best, despite a situation of generalized chaos and perennial alarmism.

In addition, respondents also praised the regional governments' crisis management, particularly complimenting the iron-fisted management of the Campania region governor, Vincenzo De Luca, who, in the opinion of the respondents, has stood up well to the threats of the virus. Less idyllic words were spoken about governmental crisis management in the Lombardy region, where the remarkably high number of infections and an initial inability to manage the spread of the virus made the implementation of containment measures particularly difficult. The governor of the Lombardy region slowed down the regional response in an attempt to not slow down regional productivity, so as not to compromise the local economy. This choice by the governor of Lombardy was deemed by our respondents to be extremely dangerous.

Even though there were many positive reactions to the management of the pandemic, negative opinions about how the situation was managed mostly referred to the inability to manage the procedures for prevention, control, and detection of the COVID-19 infection. Specifically problematic were the management of molecular swabs, the system for tracking infections, and the lack of preparation for dealing with such an unknown enemy.

The blame for the mismanagement was laid at the door of cuts in healthcare funds, which did not allow a more aggressive response. But blame for the failures of management also lies in the inability to give direct responsibility to citizens, who must play their part in helping the system. Of course, this also involves poor public and institutional communication capable of conveying the true extent of the pandemic in progress.

The COVID-19 emergency has been used by politicians for electoral purposes, since even actions that turned out to be due for the collective good have often been used as propaganda warhorses, sometimes even in opposition to the public good. In the words of one respondent, there was an "inability for the political parties to collaborate for the good of the citizens. Too much political show; too little concrete action; economic crisis; abuse of power in decision-making."

The difficulties involved in dealing with an unknown phenomenon has been the subject of no little conflict between the state and the regions in Italy. Obviously, the aid system and dedicated funds are perceived as scarce, but the idea that no one is left behind pervades public opinion and lies directly in the communicative capacity that has seen the prime minister as the direct communicator of each of the harder decisions made at the most delicate moments in the management of the pandemic. All of this was done in the face of a continuous infodemic of fake news and misinformation that continuously threatened the stability of institutional crisis management. The respondents complained about the slowness of some measures and of the arrival of support, especially economic support; they complained about the inconsistency of some measures; they complained about the lack of scope and the absence of targeted controls; they complained about the use of numbers and estimates in alarmist communications; and, above all, they complained about the lack of universality of the measures and supports that have increased gaps and inequalities, affecting some groups at the expense of others.

At a general level, what emerges from the comments of the sample is an appreciation for the work of the health actors and a dissatisfaction that was generated, by contrast, at the level of political management and socio-economic impacts. This was the case precisely because, while the work of healthcare professionals is visible and tangible, the impacts of COVID-19 on the economy and society have not yet been fully shown. Therefore, evaluating the effectiveness of the containment and support measures is even more complicated, leaving much room for ambiguity and uncertainty.

When presented with a final opportunity to leave general suggestions and impressions, our sample did not limit itself to expressing of analyses of the situation, but rather included comments that explain all the issues raised:

- "Twenty years of cuts to public health in favor of the private sector, and the fact that healthcare is no longer the responsibility of the state but has been delegated to the regions, has presented the bill in terms of thousands of lives lost. As far as the current government is concerned, there has never been a real lockdown of the production districts in the regions most affected: individual 'disobedient' citizens have been blamed, but the irresponsible ones have been others. The lack of universal support measures in the face of a socioeconomic crisis will be long-term risks having a very serious impact on large sections of the population, especially in the South. What is needed is a serious citizenship income and not spotty and fragmentary measures that do not reach the most disadvantaged. We need to regularize the status of migrants and guarantee their access to healthcare. We need to invest in schools to return to teaching in person, with concern for the safety of those who work there. We do not need to be treated as subjects; we need health workers to work in safety, hospitals to be protected, preventive and territorial medicine to work, and to track the contagion in order to contain it."

- "Despite the delays and mistakes made by the government in the course of its work, the parliamentary institutions – contrary to the political forces of opposition – have done their best to address and resolve the problems posed by the pandemic. It is an unprecedented pandemic, in that it has struck ubiquitously worldwide, and as such has found all the countries of the world and many of their leaders unprepared. Italy, having been the first country to be affected after China, from where COVID-19 spread, has in fact had to lead the way at the European level, and not only there. Prime ministers such as Johnson in England and presidents such as Trump in the USA have completely underestimated the pandemic in progress, unwittingly causing its spread, with very high costs in human lives and in economic-financial terms. At a global level, the virus does not know people's opinions; it only needs to infect them in order to survive and multiply. Whoever has not been able to consider this simple evidence is stupid and incompetent, and should therefore be removed from any position of political, economic, public health and, last but not least, scientific responsibility. (One for all, Luca Montagnier)."

• "I would like to say that I do not think it was possible to react optimally, especially because of the very tight timeframe in which it was necessary to intervene in a new situation. I believe that the institutional intervention went in the right direction to limit the spread of the virus. What was missing were measures more contextualized to the local reality of the emergency and a broader vision that would take the opportunity to resolve existing problems (i.e., not returning to normality, but creating a new normality that was better than the previous one). In addition, we need to build a proper strategy now, that will allow us to intervene in the future in a better way, should similar emergencies happen again."

• "I think the government faced the biggest emergency in our history. But some serious mistakes were made in good faith. The first was to wait until 12 March to impose a quarantine on the country. That was too late. Another mistake was not closing the borders in January. Closing only flights from China was useless because people were entering our country from other countries and bringing the virus into Italy. Other mistakes were made in regions such as Lombardy, where the regional government was disastrous with errors that led to the spread of the contagion like wildfire (such as taking the sick to the RSAs, or not swabbing everyone to find out who was infected but asymptomatic). The government treated Lombardy and the other regions in the same way. This was a mistake. Furthermore, allowing 30,000 people to take trains from the North at night while the draft decree was being drawn up was a grave mistake. I would like to know whether those who allowed the draft to come out paid for it. The stations should have been closed and no one allowed to get off the trains in the South. These people have brought contagion to the South."

• "We note the strong commitment of the government in the management of the emergency, but since no one at the government level has ever made any crisis communication plan or health management of the emergency, with COVID-19 they had to remedy this problem quickly and with difficulty, during the emergency itself. This revealed the total disinterest of the state towards the public good, along with corruption and the full responsibility of the institutions, the bad faith of the representatives of the institutions, and of the hospitals and civil protection agencies. The presidents of the regions (especially Liguria, Veneto, Lombardy, and Calabria) demonstrated total inadequacy, lack of responsibility, and racism (e.g., "the Chinese eat rats," and "If the virus becomes less virulent then it has been created in a laboratory.") The president of Liguria, Toti, refused to wear a mask when going out in public or meeting journalists; the president of Calabria ordered the complete opening of all commercial establishments in defiance of institutional provisions. COVID-19 has become (unfortunately) the means of trying to increase Italy's importance in Europe. Politicians of both the majority and the opposition parties have used the emergency as a means of campaigning."

- "The unpreparedness of the health system, which has been continually impoverished by wicked political decisions, and the underestimation of the national government, have put the country at serious risk. These conditions have seriously jeopardized optimal containment of the virus and, above all, have not prevented the excessive number of deaths. It is still incomprehensible why these factors were more radical in some regions than in others. Unfortunately, science has also shown many limitations in approaching the disease and, therefore, the most suitable treatments: There have been many contradictory positions, except on the application of the only remedy that has existed for centuries: confinement. Even the management of the measures to support the economy show a country that is highly inadequate, revealing a past of political short-sightedness, the irreparable consequences of which we have seen in this pandemic."

- "Institutional communication has been inconsistent and lacking in substance. The link between the state and the regions has not been well managed. Children, adolescents, and young people have been forgotten. Responsibility has been shifted to the citizens and not to the institutions. And there is no serious system for tracking and testing."

- "I believe that a situation has arisen that is difficult for the political forces in our country to understand. However, after an initial phase of disorientation and some mistakes, everything possible was done, knowing that we were groping in the dark and navigating by sight. Probably the greatest difficulty, in terms of institutional management, was not caused by the virus but by opposition figures who were seeking support among the less affluent sectors of the population. But even in this case, the emergency as a whole was dealt with well."

- "Immediate measures were needed, especially toward the hotbeds. Vehicles had to be stopped immediately, or adequate security measures put in place, to prevent the red zone from widening. The clash between the political factions over the decrees created confusion among the population, generating disorder and criticism of the decrees and of Conte. In

times of emergency it is necessary to remain united and trust the government's decisions, without the polemics that generate discouragement."
- "Management of the pandemic was completely entrusted to the 'sense of responsibility of citizens.' There was an absence of real measures to trace COVID; completely inadequate swabs; statistics based on partial and inaccurate data (which were, therefore, misleading); a lockdown motivated more by the inadequacy of the health system (and therefore the fear of collapse) than by fear of the virus itself; and indiscriminate containment without taking into account factors such as age, previous illnesses, weaker subjects, and the geographical spread of the virus. Extremely rhetorical political and media narration (#itwillgoallwell, #stayathome, #alsoyouhelpus) seems to have ignored the real (economic and social) consequences of the pandemic. It was a difficult trade-off, a complicated situation. Choosing between human lives and economic balance (which sometimes translates into human lives) is a major responsibility. The world did not expect such a black swan; Italy was unlucky to be among the first countries in Europe. I do not agree with all the choices made, but I admire our leaders for the serious responsibility they have taken in this period of uncertainty on the part of science (we did not know how the virus would behave). The case of governments that have chosen to stick their heads in the sand is different."

All the comments we reviewed dwell on critical and less critical aspects of the crisis, on praise and recrimination, and on conscious and well-justified communication. However, to have collected so many articulate comments is not something we could have anticipated. At this point in our web survey, the participants had already responded to about 50 individual questions and sets of questions. It is more unique than rare, in such a study, to find respondents able to articulate such deep thoughts, giving insight into what was happening at the level of institutional management in Italy.

One truly unique feature of the study is the ability to report the respondents' own words to the reader, without mediation, including the most passionate messages left by our respondents. This is truly exceptional. At another moment in time, it is unlikely that respondents (especially online) would have devoted such a great deal of time and cognitive resources to provide a heartfelt and considered response to our web survey.

Suggestions for policy, practice, and research

The third part of the present chapter is dedicated to strengths and weaknesses in managing the crisis, starting from the individual reflections provided by our sample and trying to systematize these insights into suggestions for policies and practices that will enhance people's resilience.

Suggestions for policy and practice

This study has several suggestions for policy and practice concerning future risk communication, public management, and emergency management.

In risk communication, most of our respondents believe that certain age groups (such as those older than 60) were more at risk, leaving respondents with the false understanding that younger people were immune, or that they would at least face fewer consequences if infected. In disaster science, this phenomenon is called a *moral hazard*. A moral hazard occurs when people take a risk greater than what they would otherwise have done, and they do this precisely because they feel they will not face any consequences. People younger than 60 years old may have felt less threatened and, therefore, may have undertaken more risky behaviors, thus contributing to the spread of the virus. Italy has not experienced a shift in the age group most affected by the virus. According to the National Institute of Public Health (Istituto Superiore della Salute) in August 2020, the average age of those affected by the coronavirus was 35 years old, in comparison to 60 years old at the beginning of the pandemic (Il Mattino, 2020). Data from the National Institute of Public Health shows that in August 2020, 14.8% of affected people were in the age group 0–18 years. Previously, only 1% of this age group was affected by COVID-19. So, in August 2020, 56% of those who tested positive for the virus were between 18 and 50 years old. The takeaway here is that communication with the general public needs to consider the unintended consequences of delivering messages that can be misleading.

Misleading communication about risk and a staged reopening of the country in Phase 2 made respondents feel less responsible for others' safety. People started engaging in more risky behaviors, supported by a false sense of security. By way of example, our study found that people (mainly women, young people, and families with children living in the South and the islands) were adopting fewer protective behaviors. This group of people seemed to be more ontologically secure, likely due to the fact that they were geographically more distant from the threat. Their social and physical environment did not change much, compared to somebody living in the red zone. Initially, COVID-19 primarily affected the northern regions. And, if we look at some of the respondents' characteristics for this cluster of people, we see that they belong to a group with low socioeconomic status, meaning that they are also less educated. This suggests that they might have had some problems in understanding the messages being conveyed.

Another important takeaway has to do with the importance of issuing communications that do not lead people to think that they are powerless, which would induce them to assume a fatalistic perspective. Of course, the feeling of being powerless might also be related to aspects of daily life that impact people's identity. People in this group tend to be self-employed, fatalistic, divorced or single and to have a perceived good understanding of the pandemic and, therefore, a high level of adherence to normative behaviors. Self-employed people might face a challenge in making the trade-off between believing that they are at high risk and avoiding taking responsibility for themselves and others, as well as in adopting a fatalistic perspective. For instance, self-employed people might have no choice but to take public transportation or leave their homes to go to work. Self-employed people might also find the adoption of protective behaviors to contrast with their identity (for example, as restaurant owners), and this could make it difficult for them to justify these measures to themselves. Another important thing to consider is that people who adopt fewer protective behaviors are also embedded in less-dense networks. In other words, fatalists and individualists tend to have fewer relationships (they tend to be divorced or single), and therefore the social control pressure is reduced for them.

We have seen that the credibility and trustworthiness of the message need to be combined with the trustworthiness of the source. Thus, conflicts between the central and regional levels of government should be avoided.

Another recommendation for policy and practice is to avoid surprise situations that generate panic in the public and might also generate distrust in the government. For instance, the leaking of information during Phase 1 about the imminent lockdown of the entire country triggered many Italians with family in Italy's southern regions to try to reconnect with their family of origin. Many judged the behavior of these people as a form of panic. In disaster literature, panic is treated as a myth because people rarely panic to the point of showing antisocial behavior. Leaving the northern part of Italy to reconnect with their southern families might be considered antisocial behavior, since everyone should be responsible for themselves and others in the effort to contain the pandemic. However, this form of antisocial behavior is a sort of self-fulfilling prophecy. People did not need to know about the imminent lockdown, as that information might cause panic. People hearing the sudden news must have felt trapped, powerless, and individually isolated (Quarantelli, 2001). Blaming the people affected by a disaster for panicking is a way to shift the attention onto the victims rather than taking full responsibility for managing that disaster or sharing the responsibility for solving the disaster by engaging people in the response. This study demonstrated that people tend to have cooperative and prosocial behavior when they are made aware of what is happening, what is expected from them, and the reasons behind specific choices. The pro-sociality of behavior results from the fact that people feel responsible for themselves and for others. They also tend to be more forgiving when their government faces natural disasters.

Suggestions for research

Here we offer some epistemological and methodological reflections regarding the future of social science research after the COVID-19 crisis, drawn from our research experience. The COVID-19 pandemic is undoubtedly having a profound impact on all aspects of everyday life in current society; moreover, its aftermath will be felt for many years to come in almost all fields of human activity: the economy, health, politics, science, media, and more. Of course, social science research is no exception.

The pandemic will likely lead to a cut in public spending on education, science, and research. Moreover, analysts and scholars believe that this predicted decrease in financial resources will exacerbate the gap and the problem of inequality in access to education and research (Bania & Dubey, 2020; IIEP & UNESCO, 2020; Schleicher, 2020).

One of the most substantial impacts of the pandemic on social scientists' activities has been having to cancel or postpone all opportunities for live discussions, such as conferences and seminars (Schleicher, 2020). These forms of academic and scientific debate are now held exclusively online, due to mobility restrictions. The absence of these forms of cultural and scientific exchange "may jeopardize the efficiency and innovation of research in the future" (Bania & Dubey, 2020).

From the perspective of social science research methodology, COVID-19 has had a controversial impact on the way research is designed and conducted. But make no mistake: There have also been positive effects that must be acknowledged.

On the one hand, the pandemic has made it practically impossible for social scientists to implement all those forms of data collection that rely on communication and relationships among people. This has been detrimental to qualitative approaches, whose most common research techniques, interviews, focus groups, and observation are based on the interactions between subjects within a defined social and cultural context. However, research adopting a quantitative approach, especially the survey, has also felt the pandemic's effects. In the last year, people have been overwhelmed by a huge number of research projects collecting data with online questionnaires. And, sad to say, not all of these research projects have had high quality standards. These impacts of the pandemic on social science research, therefore, raise legitimate questions about the validity and quality of the research being conducted in recent months.

On the other hand, the COVID-19 emergency has accelerated the consolidation of new methodological trends within social science research, especially by boosting the adoption of digital methods. This is because, first of all, some characteristics of the main object of social science research – individuals – have changed. Restrictions on mobility have forced people to stay at home, in a real "always connected" condition. This, on the one hand, has led people to produce data almost continuously with their online activities, posts, comments, likes, tweets (i.e., big data). On the other hand, the pandemic has made individuals more available for research carried out with quantitative or qualitative digital methods (i.e., small data).

The research presented in this book demonstrates what has just been said: In a pre-COVID era, it would not have been possible to reach a large sample like ours in such a short time. Other research carried out in Italy, using the same survey approach, has also achieved very high sample sizes (Lombardo & Mauceri, 2020). During the pandemic, social science both rediscovered traditional methods that had never been mainstream, such as network analysis or content analysis, and experimented with new solutions, like topic modeling, machine learning, and digital methods (Molteni & Airoldi, 2018; Veltri, 2017).

Social science research should play a crucial role in understanding epidemics like COVID-19 by using digital methods to explore and interpret different social phenomena during and after these critical times. However, social science researchers need to address and overcome the challenges that emerge from researching with digital methods, while making the most of their advantages. Researchers need appropriate epistemological, methodological, and technical expertise and skills, along with efficient procedures to test the reliability, validity, and quality of the research tools they are using (Bania & Dubey, 2020). The future of social science research lies in its capacity to hybridize itself with digital technology without losing its nature and its distinctive traits. In conclusion, it can be said that COVID paradoxically and ironically proved to be a great opportunity for social science research to embark on a path of theoretical reflection and methodological innovation.

The enormous production of books, articles, and publications of various kinds inspired by COVID-19 testify that the applied social sciences still have a lot to say and a lot to propose to improve our societies, helping policy-makers to develop shared and accepted strategies that may lead to an inclusive and sustainable recovery after the pandemic is over. Moreover, social science research may ensure that the voices of those communities in trouble are heard and discussed, and that stakeholders can be involved in decisions that affect them. Social science is, then, regaining a central role within the public debate on society that seems to have been taking place over the last few decades.

References

Bania, J., & Dubey, R. (2020, October 28). *The Covid-19 pandemic and social science (qualitative) research: An epistemological analysis.* www.guninetwork.org/report/covid-19-pandemic-and-social-science-qualitative-research-epistemological-analysis

Il Mattino. (2020, August 18). www.ilmattino.it/primopiano/sanita/coronavirus_iss_eta_media_contagiati_35_anni-5410982.html

International Institute for Educational Planning & United Nations Educational, Scientific and Cultural Organization. (2020, April 7). *What price will education pay for COVID-19?* www.iiep.unesco.org/en/what-price-will-education-pay-covid-19-13366

Lombardo, C., & Mauceri, S. (2020). *La società catastrofica: Vita e relazioni sociali ai tempi dell'emergenza Covid-19.* FrancoAngeli.

Molteni, F., & Airoldi, M. (2018). Integrare survey e big data nella pratica della ricerca, *Sociologia e Ricerca Sociale, 116*, 103–115. https://doi.org/10.3280/SR2018-116009

Quarantelli, E. L. (2001). *The sociology of panic.* http://udspace.udel.edu/handle/19716/308

Schleicher, A. (2020). *The impact of COVID-19 on education: Insights from Education at a Glance 2020.* Organization for Economic Co-operation and Development. www.oecd.org/education/the-impact-of-covid-19-on-education-insights-education-at-a-glance-2020.pdf

Veltri, G. A. (2017, April 17). Big data is not only about data: The two cultures of modelling. *Big Data & Society, 4*(1). https://doi.org/10.1177/2053951717703997

Index